Writing in the Teaching and Learning of Mathematics

The photo of John Meier was taken by Tricia Thorme.

Library of Congress Catalog Card Number 98-86032

ISBN 0-88385-158-X

Printed in the United States of America

Current Printing (last digit):
10 9 8 7 6 5 4 3 2

Writing in the Teaching and Learning of Mathematics

John Meier
Thomas Rishel

Published by
THE MATHEMATICAL ASSOCIATION OF AMERICA

The MAA Notes Series, started in 1982, addresses a broad range of topics and themes of interest to all who are involved with undergraduate mathematics. The volumes in this series are readable, informative, and useful, and help the mathematical community keep up with developments of importance to mathematics.

MAA Notes

14. Mathematical Writing, by *Donald E. Knuth, Tracy Larrabee, and Paul M. Roberts.*
16. Using Writing to Teach Mathematics, *Andrew Sterrett,* Editor.
17. Priming the Calculus Pump: Innovations and Resources, Committee on Calculus Reform and the First Two Years, a subcomittee of the Committee on the Undergraduate Program in Mathematics, *Thomas W. Tucker,* Editor.
18. Models for Undergraduate Research in Mathematics, *Lester Senechal,* Editor.
19. Visualization in Teaching and Learning Mathematics, Committee on Computers in Mathematics Education, *Steve Cunningham and Walter S. Zimmermann,* Editors.
20. The Laboratory Approach to Teaching Calculus, *L. Carl Leinbach et al.,* Editors.
21. Perspectives on Contemporary Statistics, *David C. Hoaglin and David S. Moore,* Editors.
22. Heeding the Call for Change: Suggestions for Curricular Action, *Lynn A. Steen,* Editor.
24. Symbolic Computation in Undergraduate Mathematics Education, *Zaven A. Karian,* Editor.
25. The Concept of Function: Aspects of Epistemology and Pedagogy, *Guershon Harel and Ed Dubinsky,* Editors.
26. Statistics for the Twenty-First Century, *Florence and Sheldon Gordon,* Editors.
27. Resources for Calculus Collection, Volume 1: Learning by Discovery: A Lab Manual for Calculus, *Anita E. Solow,* Editor.
28. Resources for Calculus Collection, Volume 2: Calculus Problems for a New Century, *Robert Fraga,* Editor.
29. Resources for Calculus Collection, Volume 3: Applications of Calculus, *Philip Straffin,* Editor.
30. Resources for Calculus Collection, Volume 4: Problems for Student Investigation, *Michael B. Jackson and John R. Ramsay,* Editors.
31. Resources for Calculus Collection, Volume 5: Readings for Calculus, *Underwood Dudley,* Editor.
32. Essays in Humanistic Mathematics, *Alvin White,* Editor.
33. Research Issues in Undergraduate Mathematics Learning: Preliminary Analyses and Results, *James J. Kaput and Ed Dubinsky,* Editors.
34. In Eves' Circles, *Joby Milo Anthony,* Editor.
35. You're the Professor, What Next? Ideas and Resources for Preparing College Teachers, The Committee on Preparation for College Teaching, *Bettye Anne Case,* Editor.
36. Preparing for a New Calculus: Conference Proceedings, *Anita E. Solow,* Editor.
37. A Practical Guide to Cooperative Learning in Collegiate Mathematics, *Nancy L. Hagelgans, Barbara E. Reynolds, SDS, Keith Schwingendorf, Draga Vidakovic, Ed Dubinsky, Mazen Shahin, G. Joseph Wimbish, Jr.*
38. Models That Work: Case Studies in Effective Undergraduate Mathematics Programs, *Alan C. Tucker,* Editor.
39. Calculus: The Dynamics of Change, CUPM Subcommittee on Calculus Reform and the First Two Years, *A. Wayne Roberts,* Editor.
40. Vita Mathematica: Historical Research and Integration with Teaching, *Ronald Calinger,* Editor.

41. Geometry Turned On: Dynamic Software in Learning, Teaching, and Research, *James R. King and Doris Schattschneider,* Editors.
42. Resources for Teaching Linear Algebra, *David Carlson, Charles R. Johnson, David C. Lay, A. Duane Porter, Ann E. Watkins, William Watkins,* Editors.
43. Student Assessment in Calculus: A Report of the NSF Working Group on Assessment in Calculus, *Alan Schoenfeld,* Editor.
44. Readings in Cooperative Learning for Undergraduate Mathematics, *Ed Dubinsky, David Mathews, and Barbara E. Reynolds,* Editors.
45. Confronting the Core Curriculum: Considering Change in the Undergraduate Mathematics Major, *John A. Dossey,* Editor.
46. Women in Mathematics: Scaling the Heights, *Deborah Nolan,* Editor.
47. Exemplary Programs in Introductory College Mathematics: Innovative Programs Using Technology, *Susan Lenker,* Editor.
48. Writing in the Teaching and Learning of Mathematics, *John Meier and Thomas Rishel.*
49. Assessment Practices in Undergraduate Mathematics, *Bonnie Gold,* Editor.
50. Revolutions in Differential Equations: Exploring ODEs with Modern Technology, *Michael J. Kallaher,* Editor.
51. Using History to Teach Mathematics: An International Perspective, *Victor J. Katz,* Editor.
52. Teaching Statistics: Resources for Undergraduate Instructors, *Thomas L. Moore,* Editor.
53. Geometry at Work: Papers in Applied Geometry, *Catherine A. Gorini,* Editor.
54. Teaching First: A Guide for New Mathematicians, *Thomas W. Rishel.*
55. Cooperative Learning in Undergraduate Mathematics: Issues That Matter and Strategies That Work, *Elizabeth C. Rogers, Barbara E. Reynolds, Neil A. Davidson, and Anthony D. Thomas,* Editors.
56. Changing Calculus: A Report on Evaluation Efforts and National Impact from 1988 to 1998, *Susan L. Ganter.*
57. Learning to Teach and Teaching to Learn Mathematics: Resources for Professional Development, *Matthew Delong and Dale Winter.*
58. Fractals, Graphics, and Mathematics Education, Benoit Mandelbrot and Michael Frame, Editors.
59. Linear Algebra Gems: Assets for Undergraduate Mathematics, *David Carlson, Charles R. Johnson, David C. Lay, and A. Duane Porter,* Editors.
60. Innovations in Teaching Abstract Algebra, *Allen C. Hibbard and Ellen J. Maycock,* Editors.
61. Changing Core Mathematics, *Chris Arney and Donald Small,* Editors.
62. Achieving Quantitative Literacy: An Urgent Challenge for Higher Education, *Lynn Arthur Steen.*
63. Women Who Love Mathematics: A Sourcebook of Significant Writings, *Miriam Cooney,* Editor.
64. Leading the Mathematical Sciences Department: A Resource for Chairs, *Tina H. Straley, Marcia P. Sward, and Jon W. Scott,* Editors.
65. Innovations in Teaching Statistics, *Joan B. Garfield,* Editor.
66. Mathematics in Service to the Community: Concepts and models for service-learning in the mathematical sciences, *Charles R. Hadlock,* Editor.
67. Innovative Approaches to Undergraduate Mathematics Courses Beyond Calculus, *Richard J. Maher,* Editor.
68. From Calculus to Computers: Using the last 200 years of mathematics history in the classroom, *Amy Shell-Gellasch and Dick Jardine,* Editors.
69. A Fresh Start for Collegiate Mathematics: Rethinking the Courses below Calculus, *Nancy Baxter Hastings,* Editor.

MAA Service Center
P.O. Box 91112
Washington, DC 20090-1112
1-800-331-1MAA FAX: 1-301-206-9789

Contents

Introduction

This is a text primarily about how to create and incorporate effective writing assignments throughout the mathematics curriculum. Mixed in with our treatment of *how* to create such assignments is a discussion of *why* we use writing assignments in our courses; this leads to a discussion of related strategies (such as student classroom presentations) which achieve similar results. So while this is a text on what some have termed "writing to learn mathematics," it is more generally an argument for engaging students in a dialogue about the mathematics they are trying to learn.

We have two main audiences in mind for this text. One is teachers who are interested in using writing assignments in their classes, but who don't have extensive experience with writing assignments. The other audience is faculty who have used writing but would like to further reflect on their methodology. We believe that a number of our comments, especially later in the text, will help instructors think critically about the assignments they are using and the goals they hope to accomplish with them.

To help give some impression of what is to follow, we quickly outline the four sections of this book.

When we conduct workshops on writing as a pedagogical tool in mathematics, we generally hear a lot of the same comments or questions:

"I'm a mathematician, not a writing instructor!"

"How do you grade an essay?"

"How does one find or come up with good writing assignments?"

These are important questions for anyone who is considering using writing assignments in their classes. We hope that our first section, **First Steps**, will partly answer these kinds of questions. We have tried to give concrete advice, with a large number of examples, in order that an instructor can successfully begin to use writing in their classroom. **First Steps** concludes with some comments on grading. While this is intended to be a section for "beginners," we hope our comments and examples will prove useful to "veterans" of the writing across the curriculum movement.

In **Listening to Others** we introduce some of the terminology and techniques used by our colleagues in writing departments to improve student writing and to improve the effectiveness of the writing experience as a pedagogical instrument. If we want to get better essays from our students, it's a good idea to talk to the people who have been struggling with that question for decades. After a general discussion we focus on a few ideas we find particularly helpful in teaching mathematics: prewriting, process writing, audience and narrative. We conclude the section with a similar discussion of taxonomies used in cognitive science, and how thinking about the issues they raise can improve the courses we teach.

It is not until our third section that we turn to what is usually the first topic many people think of when they think about writing in mathematics: **Major Projects**. Putting together a successful major project requires a lot of work from both the student and the instructor. In this section we offer numerous examples of projects we have used, along with advice on using individual conferences and classroom presentations to get the most from these projects.

In the final section, **Narrating Mathematics**, we expand on an argument that is implicit throughout this text: Writing, speaking and thinking are all intertwined, no matter what the field of endeavor. By having students communicate their mathematical ideas in narrative writing and speaking, the students will begin to better understand the material. Further, students of mathematics, like students in general, take naturally to story methods. Even so, narration is often remarkably absent from mathematics and mathematics education. The use of writing assignments fairly cries out for the introduction of narrative techniques, especially those stories and talks that the students create for themselves in order to better understand the material. By asking good questions and critiquing student manuscripts in an open, yet rigorous, manner, instructors can get students at any level of ability and background to a deeper awareness of the beauty and power of mathematics.

The process of writing this book has helped us to sharpen some of our ideas and to change others. We think it has made us better teachers and more knowledgeable users of writing in the classroom. Because our own ideas evolved even after we thought we "knew what we were doing," we don't want to be prescriptive, claiming that "if you do this, it will work." Still, we do believe that sensible, well-formed writing assignments which are consistent with the pedagogical style and goals of a course will set up the conditions for a dialogue with students that can lead to a real understanding of the nature of the mathematics being studied.

Before turning to our text, we'd like to make one final point related to the phrase "consistent with the pedagogical style" in the last sentence above. The entire discussion in this book occurs in the context of our own pedagogical preferences, which may not be everyone's. While we find certain methodologies quite effective, they will not be appropriate for all teachers. In order for this to be a useful text, we need to have critical readers. Thus early chapters conclude with questions we hope that you will consider, either by writing out answers or just pausing and thinking about them for a few moments. If writing is to be an effective pedagogical tool for you, you must design assignments and respond to student solutions in a manner consistent with your teaching style; simply reading about what we do in our classes won't help you in yours.

And finally, to quote from ourselves in Chapter 14:

Mathematics is no different from the rest of experience; it is a topic which we store in our heads as a narrative—a story about what we know, what we don't know, and what we wish to know. It's a story we must share with other people to see if we've got it right, to fill in the gaps, to make it grow. We must encourage our students, and ourselves, to share this narrative that we each have constructed and will construct. "Storying, encountering the world, and understanding it contextually, by shaping ideas, facts, experience itself into stories," says

Gordon Wells, [Wg] “is one of the most fundamental means of making meaning: as such, it is an activity that pervades all learning.”

John Meier and Tom Rishel
Ithaca, New York
May 1998

Acknowledgements

If Ken Ross hadn't offered us a panel at the Cincinnati AMS-MAA meetings, it is unlikely that we would have ever sat down in a hotel lobby to begin designing this book.

Similarly, if our students hadn't done such good work on their writing projects, we might never have become convinced of the efficacy of using writing to teach mathematics. All the students who put up with our assignments in various classes deserve thanks. In particular we have to thank the following Cornell or Lafayette students whose work is either quoted or very directly referred to in our text; Alex Agranov, Justin Collins, Dan Elsberg, John Farley, Susan Garille, Raymond Hadinata, Max Huffman, Alison Klugherz, Todd Lang, Lillian Leung, Heather Millar, Sharon Pronchik, Julie Oldakowski, Willy Rachmady, Jack Walther, David Walton and Jessica Wolfson.

Several people helped with the production of the manuscript. Arletta Havlik had to put up with typing a manuscript that came in from two sources, on two incompatible computers, with two separate ideas of how writing and formatting should be done. She never bowed under the pressure, although she had every right to do so. In addition, Doug Alfors and Michael Inguagiato designed the diagrams and figures in the text, a task that was much more time-consuming than it seems. Comments from Barbara Reynolds and June and Richard Kraus were especially helpful.

We thank the reviewers whose comments helped us clarify our thoughts and our manuscript, and the editorial staffs of the MAA for their aid in producing the final book. Elaine Pedreira and Beverly Ruedi kept us on task, even if they couldn't always keep us on time.

Most importantly, we thank Mary Ann Malinchak Rishel, Associate Professor of Writing at Ithaca College, whose comments and insights were critical to the development of our ideas and this book. She taught us about narrative, rhetoric, cognition, audience and much of what we now know about the discipline of writing. Well before we began writing, she started us thinking critically about the use of writing to teach mathematics. Our recommendation that everyone should "stroll over to the writing department" to talk to their faculty was born out of our experience working with Mary Ann on this text.

Finally, special thanks to all the dogs and teddy bears who gave unsparingly of their love and dedication to this project.

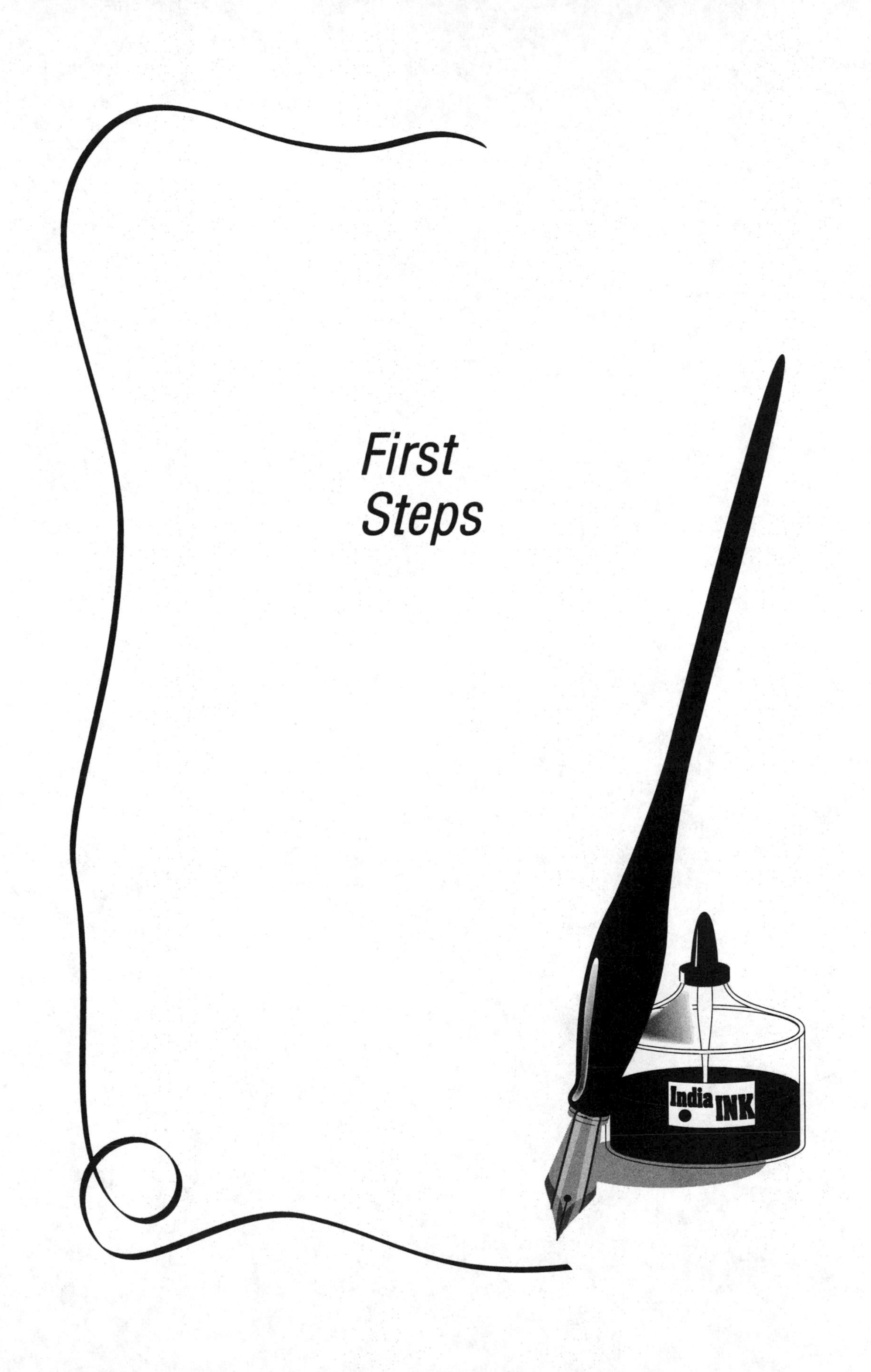

First Steps

1

Getting Started

Mathematics is often described by writing faculty as a "content discipline," which means, to be blunt about it, there's something there to write about. Yet, instructors in the field have traditionally not asked students to write about that something, opting instead for the standard mechanical exercise, "Find the roots of the following quadratic polynomial," rather than requiring a conceptual statement, such as, "Describe a situation where you saw a quadratic equation used in a nonmathematical context."

Those who always choose the rote exercise run the risk of leading students to believe that mathematics contains no concepts, no ideas, no meaning; they reduce mathematics to a bag of rules to be memorized and tricks to be remembered for exams.

It is important to ask questions: "Why did it take so many centuries to discover calculus?" "How does hyperbolic geometry really differ from Euclidean?" "Why were mathematicians so slow in accepting Joseph Fourier's methods of solving p.d.e.'s?" "Why do we add and multiply matrices the way we do?" Such assignments, while not supplanting those mechanical exercises which are so necessary for a full grasp of a mathematical topic, serve to help the student grasp the larger historical and cultural framework in which mathematics has existed for 25 centuries, and in which it will try to exist for another 25.

Tom's Narrative

I began using writing assignments in 1973, when I was teaching an upper-level statistics course at Dalhousie University in Canada. The students came from diverse backgrounds, and each seemed to have a good professional reason for taking statistics. I felt that teaching them to prove the central limit theorem, while being a somewhat useful exercise, might not be all I should do for these particular students. After teaching a rather traditional course, I assigned a final project in place of a final exam; students could select an application of statistics to their major, and if they hadn't yet chosen a major, they could "write about Canada."

The class had immediate questions, some of which I had thought about, some not:

> How many pages? Can we work together? Can I just turn in my senior thesis in biology? What's a good topic on Canada? For that matter, what's a "good topic?" Do you want references? How many?

After negotiating the above, I thought I was home free—that is, until I received the final papers. Having read the projects over for a full weekend without putting any marks on them, I came to the realization that I had not a clue as to how to grade them. All the papers seemed good, some were just "more good" than others. How could I state my opinions of their merits without hurting some feelings? I felt I had to comment on the papers . . . they had, after all, cost the students five or six weeks of work. But how could I justify any comments I might make? Worse, if someone complained about a grade, how could I convince students that my decisions were correct?

That's when I started to learn about writing in the classroom.

John's Narrative

I came to using writing assignments in my math classes 15 years after Tom. While an undergraduate at the University of Virginia, I struggled with the choice, do I become a writer or a mathematician? Finally, resolving to become a mathematician, I went to graduate school at Cornell. Still, I maintained my belief in the power of writing; after all, looking over the documents any instructor produces during a typical year, it seems there really wasn't much of a dilemma. Although many writers aren't mathematicians, just about every mathematics teacher has to produce a wealth of writing: We create exams, comment on homework, write lecture notes, and perhaps write articles or committee documents. Tom tells me he has even helped produce a set of college bylaws. Although we don't often write poetry, mathematicians are writers.

Paul Kariya, a hockey star at the University of Maine and later in the NHL, once mentioned that every time he has the puck, he wants to do something different from what he has done before. One semester early in my graduate school career I had a similar desire to "do something different" as I walked into my calculus class. Based on a vague idea that "some folks use writing assignments in math classes," I handed out my first writing assignment.

The students in this course were struggling more than usual with the notions of limit and asymptote. I devised an assignment that asked them to go to the top of Cornell's Johnson Art Museum, take a good look at Cayuga Lake stretching back toward the horizon, and then write a short essay relating this view to the concepts of limit and asymptote. I had visions of grand prose coming back, with the connections finally having been made.

The results were awful. The students had never thought of mathematics outside of problem sets, and they had no idea of how to relate the view of the lake to the mathematical concepts which were still quite fuzzy for them.

Our Common Narrative

Both our first assignments were certainly well-intentioned and salvageable.

Tom: My assignment was too unfocused; the students did not have sufficient direction to produce quality papers. This lack of focus also manifested itself in my own anxiety over grading; I wasn't clear on what I could reasonably expect from my students, and I hadn't thought about what I was looking for in their papers.

John: Part of the reason my first assignment failed is that I had given a difficult exercise without previously establishing the groundwork. In my question, I asked the students to view mathematics in a manner very different from what they were used to, and no preparatory work had been done to establish the connections the students would need to bridge the gap between "asymptotes" and "views" [Me 1]. (In Chapter 3, one we both think we could have benefited from, we will discuss ways to avoid making bad assignments, and how to salvage a good idea contained in a poorly designed problem.)

John: Although my first assignment did not have good results, I still felt intuitively that writing projects could be useful; I did not give up on using writing in mathematics courses. Remembering that Tom often used writing in his geometry class, I sent him an e-mail note asking for advice about the use of writing assignments.

Tom: I had continued to experiment with writing over the 15-year interval, so I sent a lengthy response offering some basic ideas, pitfalls to avoid, and giving examples of assignments I had used successfully in geometry. In fact, that e-mail discussion could be considered the start of this book.

John: Thinking about what Tom had written and what had occurred the previous semester, I waded into deep water once again and started assigning writing exercises in a pre-freshman summer course. Gradually, papers improved. The students' writing was better, my assignments were more focused, and I learned how to be a more effective grader.

Tom: Our discussions also became more focused; we began showing each other the impressive final projects that had been produced in upper-level courses as well as the interesting work calculus students had done. Most of my experience has been in using writing either in advanced courses or in other settings where the syllabus is very flexible, while almost all of John's teaching occurs in service courses with set syllabi. Together we have discovered that although our actual assignments differ, we both can use writing effectively in almost any mathematics course.

John: For example, when I took a job at Lafayette College, I moved to a department where computers were a required component in the calculus curriculum. I found that I had sufficient experience in using writing to enable me to create writing assignments that would enhance the effectiveness of the computer laboratory.

Now that we've told you our stories, we'd like to begin helping you put meaningful writing assignments into your classes so that your learning curve won't be as long as the ones we had.

Writing requires thinking; and thinking is what we, as teachers, want to encourage. In what follows, we hope to convince you that to get students to absorb mathematics, or any other subject, better, you need to have them think about, then write about that subject. Let students tell you their thoughts, their confusions, their half-formed ideas, their frustrations and triumphs. From this, they will understand better, and we instructors will, too, what the process of learning is all about.

Exercises/Discussion Questions

1. You've just gone to a conference about using writing in the disciplines. Some of it sounds pretty good to you, some doesn't. You'd like to try a writing assignment in your calculus class, "just to see what happens." What sort of question do you think you might ask? (Before actually *using* your question, we strongly recommend reading through the next couple of chapters.)

2. Are you an experienced user of writing in mathematics courses? If so, what short advice would you give to someone who is just starting to use writing?

3. Many people share our anxiety about grading writing. Somehow it seems less objective than grading proofs, for example. However, is it really? Make a list of things to look for when grading proofs. Are there similar things to look for in an essay?

2

Using Small Assignments

> You've gone to a conference on teaching methodology, and come back with some really great ideas, especially on writing. You just know writing will get your students to tell you what they understand about mathematics. You are ready to give them their first writing assignment! You have only one question: What will you ask them to do?

The hardest part of using writing assignments in our mathematics classes is deciding where and how to *start.* We might naturally hesitate to do so because we cannot or do not want to revise our syllabus around writing assignments. Perhaps we find the more advanced writing assignments which we have read about to be somewhat daunting: after all, our students are barely doing the work we are asking of them right now; how can we expect them to write a five-page paper on the history of calculus?

This chapter contains fairly safe, short, simple assignments to encourage writing in our classrooms.

Keep in mind that the first time we assign a writing project may well be the first time our students have been asked to write in a mathematics course. We advise minimizing the trauma (for ourselves and them) by *starting small.* Even if we plan to use longer writing assignments in our course, we still make frequent short essay assignments. In these assignments the students demonstrate the level at which they are able to communicate and what they are actually interested in. It also helps us and our students get used to writing in a mathematics course.

When reading through these sample short assignments, we ask that you consider the information you will gain about your students and how this information might help you improve your class. Our experience has shown that, when using short writing assignments, what we learn about the students and their understanding of the topic at hand is extremely useful in the day-to-day structuring of lectures, homework, and worksheets. If we want to use a longer writing assignment, such information will help us produce an appropriate project which will match our class's personality. (We will discuss this more fully in Chapter 9 on prewriting and writing.)

1. In almost any class there are short-answer questions which we can add to standard homework assignments without appreciably increasing the students' or our own workload. For example, if we are teaching a first-semester calculus class and discussing derivatives, we can certainly ask a short essay-based, but still traditional, question:

> Create a function which is continuous at $x = 0$, 2 and 4 but which is *not* differentiable at these three points. Explain why you constructed the function as you did and why it is not differentiable at these points.

The first sentence above is a nice exercise in and of itself; to solve this problem, the student will have to be a bit creative. However, it is the word "explain" in the second sentence that is key to this question; not only must students be agile in finding an answer, they must also be articulate in explaining *why* their answer is correct.

We continue to be surprised by the number of students who, having memorized our examples and seen our perfect lectures, can muddle their way to a correct answer but cannot outline the process they used or why their answer is correct. Having the students write explanations, or tell why they cannot do so, gives us a window on their understanding of the material. That, in turn, tells us how our lessons are being received and which material needs to be reinforced.

2. Having begun with simple assignments, we can now be a bit more daring with our initial questions. If our precalculus class is struggling with the formality of the concept of function, we might try to engage their imaginations by using the following assignment:

> Below is the graph of the function $f(t)$ which represents your height above the ground at time t. Write a brief explanation of what you are doing to make the graph look like it does.

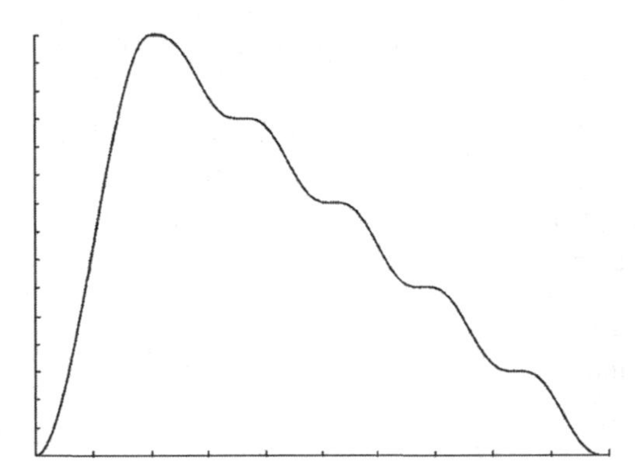

FIGURE 1

Notice that in this graph the axes have been left intentionally blank in order to allow the greatest possible latitude in student interpretations. This assignment has generated some very intriguing answers: One student's answer described herself parachuting; she first flew up in a plane, then glided slowly down. A second student's answer involved riding an elevator up to the top floor of an apartment building and then walking down the stairs to the main floor. When this question was presented at a faculty forum on assignment design, one professor explained that $f(t)$ represented his height while he was learning how to ski (he rode the chairlift up the mountain and then fell quite a bit on the way down).

Such narrative assignments are not meant to directly improve students' mathematical exposition, nor are they immediately effective in improving their problem solving skills. Rather, these sorts of assignments are designed to increase the students' comfort with a particular mathematical idea and to provide some simple projects to start having them write about mathematics.

We have used this "height above the ground" question as a starting point for further assignments meant to develop students' intuition for functions and to lead them toward understanding them from a graphical viewpoint. In this way we allow students to begin writing in their mathematics course without initially requiring them to wrestle with intricate mathematical topics. Many students are more comfortable writing a personal essay than a formal piece. Further, since our students may not be accustomed to writing for a mathematics class, asking them to work in a format in which they are comfortable will reduce some of the stress they may feel because of our "odd" assignment.

3. As part of their daily homework assignment, students can write about a specific aspect of the material they find confusing. We sometimes, for instance, require them to give a concrete example of what they find difficult; if students are very confused, this forces them to try to discover what is bothering them. Then, as a follow-up assignment, we have them write an explanation of how they overcame their confusion, once again requesting them to be specific.

For instance, a typical question for a geometry class is "What do you think *straight* should mean on a sphere?" Such a question leads to lots of classroom dialogue, usually starting with, and often ending with, "Well, what is straight *anywhere*?"

Assignments like these not only give immediate feedback on how well our class is doing; they also open avenues for dialogue. For instance, some groups of students realize they are confused on different topics and they elect to study together. Then, when the second assignment is due and they still haven't overcome their confusion, they sometimes stop by our office to ask for some help.

4. Let's assume we've reached the end of a moderately complex topic in our course, for instance, eigenvalues and eigenvectors in linear algebra. We hand out the following assignment:

> Your best friend from high school is also taking linear algebra. Because her professor has gotten behind on the assignment sheet, her class is not going to start this topic until next week. Write your friend a one-page letter explaining what you think is the most important fact or facts about eigenvalues and eigenvectors. Include at least one good example so that your friend will understand your point. Be careful not to use terminology your friend hasn't learned.

Along the same lines, we might also ask our students to create their own good problems which their friends could work on in order to motivate the study of eigenvalues and eigenvectors.

Assignments like this can produce very surprising results. Some students will do an outstanding job, others will do a workable job; meanwhile, we will learn the enormous variation in what students find important and interesting with respect to any particular topic.

The next few suggestions come from other texts as indicated below.

5. In Sterrett [St], Tim Sipka uses a variety of letter-writing assignments. He suggests having students write post-exam letters to their instructor outlining how they prepared for the exam, what their reaction is to their performance, and how they plan to prepare for the next exam. Such essays provide excellent platforms from which to discuss effective studying strategies.

Sipka also suggests having students write letters to the author(s) of their textbook. The students can focus on particular sections, indicating what is helpful and what is confusing. Reading through these letters will give good insight into how our students are using their book, while the feedback will indicate to them that we do expect them to use their text.

6. In a recent *PRIMUS* article [AA], Barbara Allen and Gwen Autin describe how they use the first five minutes of class for free writing exercises. They present their students with "prompts" to which the student has five minutes to respond. Here's a sample:

> When asked to square a binomial, many students answer that $(x+5)^2 = x^2 + 25$. Is this correct? Defend your answer.

Of course, this idea of prompted writing could be used in almost any course. Below are prompts we came up with for a linear algebra class and a real analysis class:

> Write down a 3×3 matrix $\mathcal{M}$ which has an inverse and a 3×3 matrix $\mathcal{N}$ which does not have an inverse. Explain how do you know $\mathcal{M}^{-1}$ exists and $\mathcal{N}^{-1}$ does not.
>
> Your next homework assignment will require you to prove that $\sin(x)$ is a continuous function. What will you have to prove? How might you go about this proof?

Allen and Autin describe how these prompted writing exercises focus the class while identifying areas of students' confusion. The students are preparing for what follows by writing a short answer at the start of class and reminding themselves of what they should be able to explain at this point.

7. As we get bolder, we may be willing to start our class with free writing, not focused writing. An example:

Often students have very interesting random thoughts which a highly flexible teacher can use to motivate and develop material. A rather radical example of this approach is provided by Joan Countryman [Co], who tells her students to "Write about anything as long as you're willing to let me see it." This assignment can be quite effective for some students, but it can also be too uncontrolled for other students to handle.

8. At the start of their courses, many instructors ask their students to write a "Mathematical Autobiography" discussing both positive and negative experiences they have had in their mathematics education. (See, for example, Buerk [Bk].) We think that, at first, such an assignment should be short and focused. We don't want to read pages and pages about horrible teachers in a particular student's past, or reams of boasting; what we do want is to have our students recall and reflect on specific experiences.

Reading through and commenting on our students' experiences has helped us develop a relationship with them and has given us insight into the particular paths they have taken to our classroom. Understanding their likes and dislikes allows them to be better motivated to learn the topics of the course. For example, we sometimes have a lot of complaints that "all we ever do" is memorize formulas; then when students' eyes glaze over as we begin to provide explanations as to how certain equations arise, we can remind them that they previously complained about only memorizing formulas.

If we are teaching a calculus class with a fairly rigid syllabus, short exercises like these may well be the only writing assignments we can assign. Using a few short writing assignments can still help our students learn the material, present organized solutions to problems, and approach problem-solving in a more coherent manner. Such assignments will demand solid thought from students.

Here is an example of this last: Toward the end of a third-semester calculus course, students were asked to compute the volume of the intersection of two cylinders of radius one. During the semester, the class had been using such writing assignments as "letter to a friend" exercises so that they could present in an organized manner the mathematics they were learning. The general theme of all the small writing assignments was that explanations should be clear and complete. The following student sample solution, while not a first assignment, was actually *typical* of the later work received.

We are supposed to compute the volume of the intersection of the cylinders $x^2+z^2=1$ and $y^2+z^2=1$.

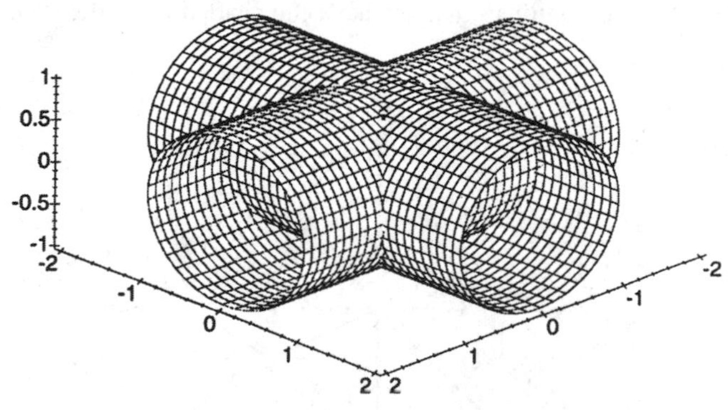

FIGURE 2

Since we are interested in only the intersection area, the part we are interested in will look like

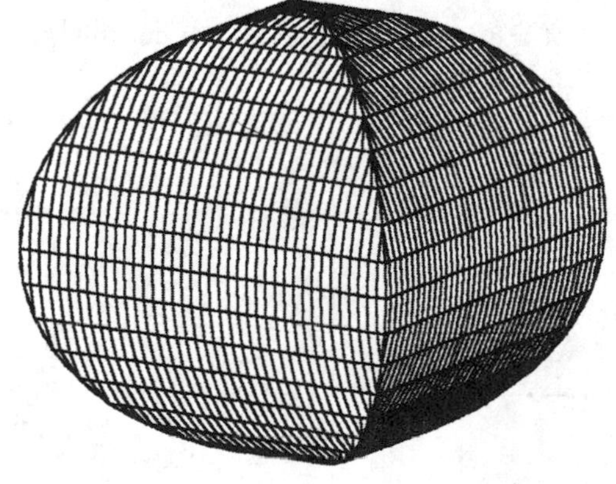

FIGURE 3

Consider the top as an umbrella with four segments and look at this object from the top.

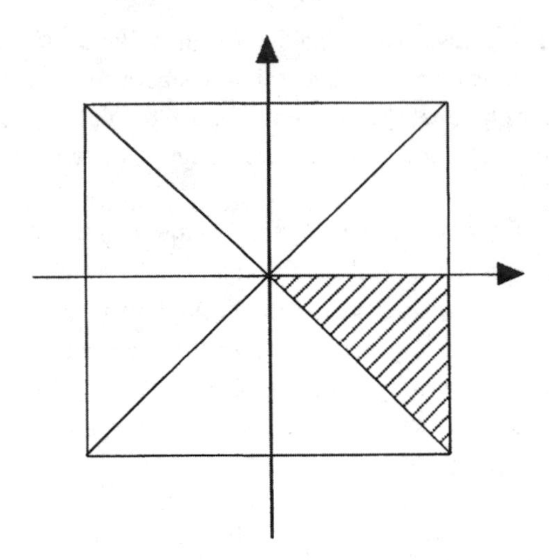

FIGURE 4

If we cut the object vertically so that we have the shaded area, then we have

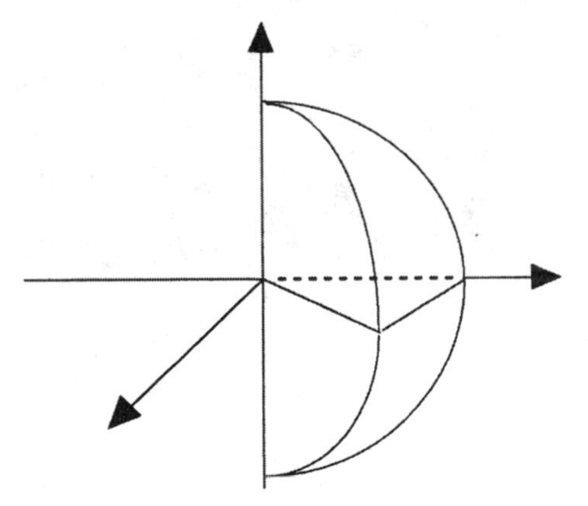

FIGURE 5

which counts for $1/8$ of the total volume. Next we cut this piece into two through the xy-plane so that it looks like

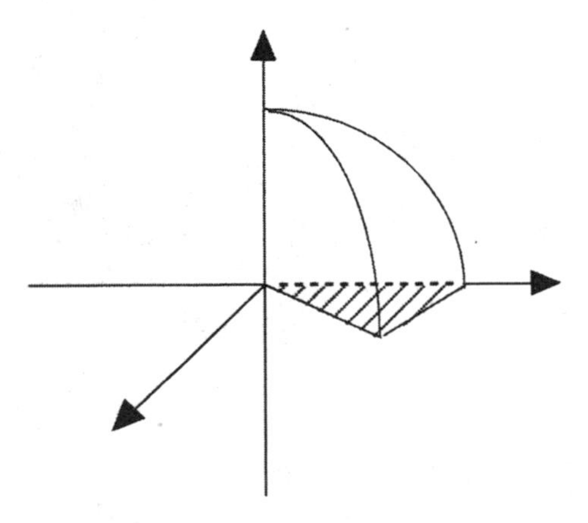

FIGURE 6

This will count for $1/16$ of the total of the initial object. Based on the shape presented, we can actually compute the volume using triple integration. The surface on top is determined by $y^2+z^2=1$, hence $z=\sqrt{1-y^2}$. Therefore, the volume of this object is

$$\int_0^1 \int_0^y \int_0^{\sqrt{1-y^2}} 1\, dz\, dx\, dy$$

$$= \int_0^1 \int_0^y \sqrt{1-y^2}\, dx\, dy = \int_0^1 y\sqrt{1-y^2}\, dy$$

$$= \frac{1}{2}\int_0^1 u^{1/2}\, du = \frac{1}{3}u^{3/2}|_0^1 = \frac{1}{3}.$$

Since this portion of the volume counts for $1/16$ of the total volume, the total volume is $16/3$.

Certainly, the short and simple writing assignments in this class helped create an atmosphere which supported and encouraged the students to present their work in a clear and organized manner. By writing about their thought processes in such problems, the students showed that they understood the individual steps behind a difficult multivariable calculus problem.

Adding a few writing assignments of the above sort does not substantially change the nature of a mathematics course, nor does it really affect the amount of material covered. What it does do is ask the students to pause and consider how they are receiving some of the important topics in the course.

Exercises/Discussion Questions

1. Look over the suggested writing assignments in this chapter. Which fits your personal teaching style the best? If you walked into class tomorrow and used this assignment, how would you expect your students to react? How would you deal with this reaction?

2. Think about the last class you taught. What was a crucial concept you presented in that class? Create one or two short writing assignments which will help your students focus on issues relating to that topic.

3. Use the next five minutes to answer the following "prompt":

> Pick one specific class that you're teaching. Would this group of students respond well to prompting questions? Would prompts fit well with the material? Why or why not?

(Five minutes is actually a long span of time, so you can pause for a couple of minutes to consider your answer before writing.)

4. Consider the student's solution to the volume of intersection example given at the end of this chapter. In what ways does this show that incorporating writing in a mathematics class alters the course?

Coda A — Some Small Assignments

Here we collect a few specific examples of small assignments we have used in a variety of introductory courses in order to give more indication of what sorts of questions could work.

In *any course*:

1. Your roommate has not come to class. In one or two paragraphs, describe today's lecture for him or her.

2. On a 3×5 index card, write down one question you have about the current topic.

In *algebra*:

3. Find an example (from outside the textbook) where you have seen algebra being used. Good examples often occur in sports or business, but you can find examples in lots of other interesting places. Give a clear description of your example and how you think algebra is being used in this example.

In *elementary statistics*:

4. Look in the daily newspaper for an example where someone cites a statistic. Based on the information in the article, can you determine exactly how this statistic was computed? Try to find at least two reasonable interpretations for exactly what has been computed, and write a couple of paragraphs describing the various ways the stated statistic could be interpreted. (This was suggested to us by Dick Furnas.)

In the *calculus* sequence:

5. Because of your excellent work in this class, you've been hired by the mathematics department to tutor other students who are taking calculus. Aaron walks in during your shift, and he is working on taking derivatives. He writes:

$$\frac{d}{dx}(x^3 \cdot \sin(x^2)) = 3x^2 \cdot \cos(2x).$$

How do you respond to Aaron's solution? Write out a dialogue which might occur between you and Aaron.

6. Below is a region in the plane (Figure 7). Find a good approximation to its area and write out an explanation of how you arrived at your answer. Make sure your explanation is sufficiently detailed that everyone in our class can understand it. Conclude your essay by describing how you could get an even better approximation if you had more time—and patience.

7. You've just come back from the "infinite toys store" where you purchased infinitely many building blocks. The first block is a $1 \times 1 \times 1$ cube, the second is a cube of side length $\frac{1}{2}$, the third of side length $\frac{1}{3}$, and in general the nth block has side length $\frac{1}{n}$. Explain why it is not possible to stack these cubes one on top of the other inside any room on this planet. Also explain that if you stack them in some other manner, you may well be able to get them inside of your dorm room.

8. Sit down and draw the graph of a function of two variables that has the peak of a "mountain" at the origin and which goes to zero as you move farther away from the origin; then find an explicitly defined two-variable function, like $f(x, y) = x^2 + y^2$ or $f(x, y) = \sin(xy)$, whose graph looks something like what you've sketched. Write a one-paragraph note explaining how

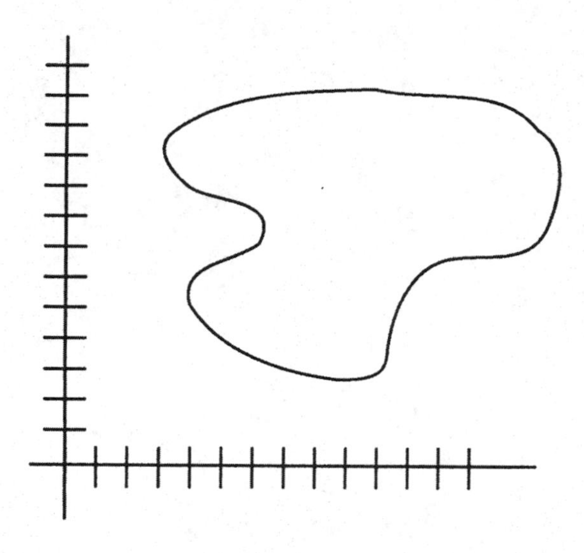

FIGURE 7

you created this function. Was your first guess perfect? If not, why was it wrong and how did you fix it?

9. Write a one-page outline describing when to use polar coordinates and when to use Cartesian coordinates.

In *linear algebra*:

10. Your friend at Rival University is having a very hard time understanding the idea of eigenvectors. Write out one more example, like the two in the book, describing how to find the eigenvectors of a specific 4×4 matrix, and what you have learned once you've found the eigenvectors. You can use any matrix you like, but make sure it's sufficiently interesting that your friend will learn something new from your example.

3

How to Salvage a Bad Assignment

At dinner at a research conference, you are sitting next to a colleague who declares: "I really don't think writing assignments do any good. Look at me, for instance. Last semester, I asked my calculus students for a five-page paper on applications of calculus, and, I tell you, the results were abominable."

There are plenty of bad assignments out in the world, just waiting to be given. For example: we have been teaching traditional calculus, and we have pointed out all the calculus-type concepts that Barrow, Newton's teacher, knew. So we say to the students, "Write a paper on who really discovered calculus." Well, we asked a good question—for a thesis topic. But any poor freshman will be totally at sea: "Where do I start? How many pages do you want? How soon do you want this paper done? How much will it count?" Those are the easy questions; the real question they have about this assignment is the one they most likely won't ask out loud: "Why didn't you think about what you wanted before you just arbitrarily gave us this assignment?" And that's also the question we'll be asking ourselves when we try to grade the "essays" we receive in response.

How do we get out of the mess we've made by asking the above tough question?

Dialogue could help—somewhat. If we gave the assignment yesterday and students are upset today, we would tell them we made a mistake, apologize profusely, promise we won't grade them on the assignment, and then ask them to participate in a discussion of "what they found out," "what they think of the question now that they've had a couple of days to think about it." (This is hypothetical; see, we'd never really do this.) Then, if we have begun to emphasize historical topics in our exposition, later in the semester we may even be able to gently bring the topic up again—by comparing Newton and Leibniz on the chain rule, for instance. In a "historical approach to calculus" course we may also be able to generate some *small* papers on the theme: "Some say that fluxions are more appropriate to physics than mathematics. Do you agree or disagree, and why?" Or, "Give one reason why you think people would call Leibniz a 'more pure mathematician' than Newton."

What we get out of student assignments is a function of what we put in. If we ask them to "write about limits," we should not be surprised when we get back "Limits are important. They are what makes calculus go." The "F" grade we give that student is one we've really earned ourselves.

A better question on limits would be the following:

We have decided that the limit concept has something to do with "approaching a point." Consider the five curves below, and talk about how the limit idea applies at "favorite points."

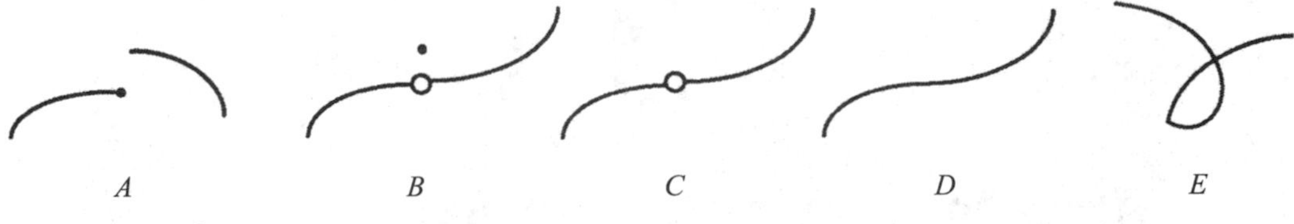

FIGURE 8

Do you see any "uniformity of approach" in any of these curves? Does it matter in which direction you approach the points? Does the word "limit" make sense in curve C, even though "a point is missing"? Is there any way to resolve the difficulty posed by the fact that curve E crosses itself?

Of course, this last assignment is much more focused than the first attempt; it is specific and concrete—and one we have used. And, because the *question* is focused, it will lead to a much more intelligent *answer*—one where both student and instructor will learn much more, and interact better, in future discussions of the topic of limit. Further, with this assignment, the instructor will actually have a chance at being able to formulate a grading scheme which leads to this interactive dialogue.

Another example: In a geometry course, we are introducing spherical geometry, and the class has been talking about properties of triangles. We bring up the idea of "straightness" on a sphere and are listening as the students question how they might define the idea of a "triangle on a sphere." The time has come to see whether the students are coping with the ideas we have introduced; let's use a writing assignment to find out what they have gleaned from the discussion.

We could ask again something simple, like "*Now* how will we define a triangle?" But, of course, that question is too vague; for instance, it doesn't even relate to whether we want the students to talk about triangles on spheres, or simply define Euclidean triangles. Instead, let's start as follows: "We have been talking about triangles on spheres. . . ." This gives the students the right to continue to "just talk." Then we would go on with "What do you (placing the student in the middle of the dialogue) think of Jim's idea that a triangle is something like 'where three geodesics come together. . . ?' What's good about his proposed definition? How might you change or sharpen it? Can you see some ways in which you might disagree with the definition, as posed?" A concrete sample of this situation is provided later on in Chapter 9.

In the above, the students' papers will be more like a dialogue with us rather than a set piece made up to give the students something to do. One small proviso: In the question, we will need to use some delicacy with regard to the mythical "Jim." If the student chosen is rather sensitive about his or her abilities, then, rather than attributing to Jim the "definition" we will soon be critiquing so heavily, we might prefer (and Jim might also like) the following less personalized approach: "One proposed 'definition' of triangle was. . . ."

Some general suggestions, which we will also return to in a later chapter:

- We try to ask the kinds of questions which afford students the possibility of responding in their own words.

- We hope to let students show what's in their mind, not what's in a book—and certainly not what someone may have told them is "the answer."
- We allow dialogue to occur before, during, and after the writing assignment.
- Before they begin, we help the students through *some* of their puzzlement, but without giving away the store.
- Sometimes we allow the writing phase to take place during class, so as to make it just another part of the ongoing dialogue between all the people in the class, not just a private, communication between all-caps US and poor little lower-case teddy or jenny.
- And after the writing, we use our comments in an ensuing dialogue with the students—not to point out how wrong they are, but to enhance the sharing of ideas and to encourage leads as to what a good, common definition might be.

As an example of further questions we can explore after the initial writing is complete, here are some sample student statements to contemplate (I have italicized words in the dialogues that you might want to ponder):

"I noticed that Jay said that triangles can get 'too large' on spheres, but not in Euclidean space. *Maybe you'd like to* tell us what you mean, Jay?"

"*One of you said* that he or she'd like her spherical triangles to come from any kind of curve on a sphere. What do you all think of that? Can you make an analogy with what might happen in Euclidean space?"

In response to a question about measuring angles: "... Oh, yeah, we haven't really talked about that... *I'm glad you brought that up, Jane. How would you cope with that* on a sphere?"

The italicized words serve as softeners for the students, making the experience of being asked questions in a classroom situation somewhat less threatening than it may be otherwise.

To summarize this chapter:

Use writing assignments, but think before you do so. Ask yourself what you really wish to accomplish with the assignment, and then give an idea of your expectations to the students. Is your question manageable?

Make sure you design questions that will generate focused and thoughtful discussion and that offer meaningful follow-up possibilities.

Let students ask you questions about what you want; set up a dialogue both before and after the actual writing takes place.

A final thought:

Quality writing assignments are made from ongoing and evolving questions. Such questions lead to a dialogue between student and instructor, helping the student to think carefully about the assignment and to grow into an awareness of the meaning of the mathematical concept.

Such a dialogue is one of the earliest and most important ways of getting students to do meaningful mathematics.

Exercises/Discussion Questions

1. Improve on the first "limits" example by rephrasing the questions. Then consider the situation you will find in the classroom after you return the students' papers. What kinds of dialogue do you think you will encounter?

2. Formulate a question about the Riemann integral which might get the students to a deeper understanding of the concept. First, pick a particular aspect of the definition you think important, then design a question which will lead to thoughtful answers. Finally, think your way through some of the dialogue you will be having in class after the students have written about the assignment.

3. Advanced students ask questions not dissimilar to those discussed in this chapter. An abstract algebra student will ask, "Where did groups come from?"; analysis students ask for a reason to study the inverse function theorem; differential geometry students want to know what the fundamental forms mean. Choose one of the above questions, or a similar topic from a course you have taken, and formulate a short response. (Yes, you may use a text, if you wish . . . and yes, this will count toward your final grade.)

Coda B. From Bad to Better Assignments

Sample 1.

Bad: Discuss continuity.

Reason: Too vague; too generic. The student will either copy statements from the text, or write down random thoughts.

Better: Which of the following three functions is continuous, which not? Give reasons for your choices.

[Reader: Insert your favorites here.]

If you were trying to make the discontinuous functions above continuous, how might you do so? Is it always possible to do that?

Sample 2.

Bad: How was calculus discovered?

Reason: This assignment is also vague, but this time the student is being asked much too much.

Better: After reading the assignment from Struik, in what sense do you agree that Newton should be credited with discovering the derivative rather than Barrow or Pascal?

Sample 3.

Somewhat Bad: "Explain how to find the roots of a quadratic equation . . ."

Reason: Not really a bad question, but it leads to only a single, prescriptive answer.

Better: Continue the above with ". . . and give one example to show why you might want to do so."

Sample 4.

Bad: State Euclid's axioms. Why are they important?

Reason: You will get writing, but you won't get thought.

Better: What was the impulse that made Euclid think that he could capture all of geometry in one short set of principles called axioms? Do you agree with his impulse?

4

The Journalistic Impulse

You're reading a book about writing across the disciplines. The author recommends that students keep journals. It sounds like such a simple thing to do; but, then again, so did using writing assignments—until you tried them.

A large number of texts on writing suggest that faculty ask students to keep a journal so as to record their impressions and insights on the topics of the course. There are positive and negative aspects to journals; we will deal with both.

Here are some cogent reasons for asking for journals:

- Students can more easily achieve the difficult task of writing in their own voice.
- They feel they have an ownership of the ideas presented.
- A journal entry is an assignment they "can cope with."
- Students have an immediate record of their own development with respect to the course and its ideas.
- The journal can be a source of interactive discussion in the classroom.
- Since writing incorporates thinking, journals can generate thought.

An imaginative, careful use of journals can be an easy way to introduce writing into a class; students have a ready venue for their ideas—a venue which always listens, without being evaluative, to any of the questions, difficulties, or conjectures that the student might be thinking about at any particular moment in the term. Meanwhile, by reading the students' entries, an instructor can get useful input into how the lectures and readings are being received. That instructor can then design future discussion, projects, and exams to address the students' needs.

In spite of all the arguments for using journals, neither of us has emphasized them in our classes, although we have used, and continue to explore, many of the techniques described above. (As you will see from the next coda, the first time he "really" taught a mathematics course with a writing component, Tom did keep a useful journal of his own in which he recorded his impressions.) If the rewards are so clear, we had to ask ourselves, why are we not assigning journals more often in our classes?

We came up with a number of reasons, some serious, some not. First, we will list them, then offer discussion:

- The instructor who asks for journals must read through a lot of "garbage"; that is, half-formed ideas and commentary.

- A journal is not an alternative to real dialogue in the classroom.
- The assignment, "keep a journal," which seems at first so simple, in actuality may lack focus.
- As to the "ownership of (mathematical) ideas," what's the benefit of ownership if it's of a wrong idea?
- With the concentration on journals, the focus of the course can shift from the mathematics to "What I feel, how I react. . . ."

Having framed the arguments, let us now discuss them.

The "garbage" idea is, of course, minimal reason for rejecting the use of journals; we have all read unfocused or disinterested writing from students, whether in the form of a diary entry or an essay for which someone won the "brilliant assignment" award from the college's writing program.

The second reason, that journal writing is not an alternative to "real dialogue," gets more to the point. We both often conduct class by first introducing a topic, next asking questions and encouraging dialogue, and only then giving a writing assignment; in this way, much of the half-formed thinking and "prewriting" has been accomplished before writing comes into play. Further, students have the opportunity to try out their conjectures on each other, as well as the instructor, in this prewriting/dialogue phase. So, in a sense, you might say that we are starting with a "spoken journal."

This "real dialogue" point also has to do with the "ownership of wrong ideas." Although we want to encourage students to pursue unusual or uncertain lines of exploration of mathematics, we also want them to have to confront objections *before* they have become so attached to completely wrong conjectures that it will be difficult for them to understand or cope with later developments in the course.

To illustrate this principle, return to the "What's your definition of a triangle?" assignment brought up in example three from the second chapter:

Sometimes students will come up with a "definition" which inadvertently allows edges of the triangle to cross each other; in a classroom discussion we instructors can construct counterexamples:

"You want THIS: 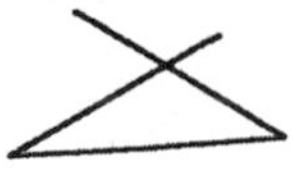 to be a triangle?"

FIGURE 9

"No, no, that's not what I meant, let's see, I'll try again...."

Meanwhile, a journal entry often has an incorrect or poorly spelled-out definition, followed by a discussion of implications of this "definition"—not irremediable, but a consideration.

Sometimes journals focus on asking students how they *feel* about mathematics. Many students say at first that they do not like the subject, based on previous courses/teachers/difficult assignments. While it is entirely fair for a student to say this, we are not inclined to dwell on this feeling. Our goal, rather, is to create a desire to learn potentially difficult, but interesting material. We prefer what happened in one of our classes on surface theory and dimension, where one of our students exclaimed, "I like this course, 'cuz there's no math in it!"

What we are saying here is not that we disbelieve in journalistic methodology. In fact, we are all for it, if properly applied. Our core problem with "just giving journals" as an assignment seems to go back to our major argument from chapter two: Assignments must have focus.

Merely saying "Keep a journal..." is not prescriptive enough for students. The instructor must continue, "...and such a journal should contain such things as transcriptions of classroom discussions, questions you have, ideas you were mulling over on the way to class, points you still don't get and want to raise, notes as to what is and is not 'working' for you, and other topics as they occur."

Also, assignment of journals cannot be a substitute for discussion. The instructor must be prepared to read student entries on a regular basis and to answer student questions as they arrive in the journals.

The instructor must also make clear to whom the diary is being written, whether and when it will be handed in, where it fits into the realm of classroom assignments ("How will we be graded on this?"), and whether it will be shared with other students.

So, let's suppose we decide to use journals in our classrooms; how would we do so?

First, we would probably start by assigning the journal for, say, only the first four weeks of the course. This would give us and the students an opportunity to "see how things are developing."

Next, we would occasionally suggest topics on which the students should concentrate. ("Tonight write about how the discussion has changed your view of four-space." Or, "Ask yourself one good question concerning what you still don't understand about the meaning of the second derivative test, and then try to formulate an answer.")

Third, we would provide feedback by reading and quickly evaluating the journals after almost every class. This evaluation might be in number format: "0" for no writing, "1" for uninteresting/unfocused writing, "2" for acceptable work. Further, each entry would receive a single, short comment, usually in question form, to stimulate further dialogue. (For some comments on what happens when grading is allowed to expand, see Buyske [By].)

If the above suggestions are followed, we believe the journal can become a useful interactive tool for maintaining student and faculty interest in the material without detracting from the original goals of the course.

Exercises/Discussion Questions

1. React to the assignment "Tell me what you don't like about mathematics." To what extent does it seem appropriate/inappropriate? If you feel it's appropriate, at what point in the class? And, if not, what is your reason for rejecting it?

2. Do you feel that journals are more appropriate in some kinds of courses than others? If so, which kinds of courses? What are your fundamental criteria for deciding that journals do not fit the other courses?

3. We tend to use assignments which grow out of classroom discussion. How does this method differ from "keeping journals," and do you have a preference of one over the other?

Coda C—Excerpts from Tom's *From Space to Geometry* Diary

Although the following is not a sample of student writing, it does give an idea of how an individual—in this case, the instructor—can use a diary as an aid in designing and evaluating a course.

31 August

The Math 150, *From Space to Geometry*, course starts today. I am not going to try to make any formal plans until I meet with the students, both to get an understanding of their abilities and also of their needs. So far only seven are registered. I will give them a list of some of the possible topics we may do, along with some of the more important "words" we will be using, e.g., curvature, distance, surface, etc. I will tell them that they will have a final project, and let them know that it will be graded partially on the basis of the writing skills they show. I will mention that there will be a number of homework projects which may include some construction of models as well as the more traditional math homeworks. I will also make some suggestions as to possible final projects, just to get them thinking in terms of what they may expect at the end of the term. If possible, I will end the day with a discussion of what the words "space" and "geometry" may mean, and why I chose them for the title of this course.

1 September

There are five students; one college scholar (art history, mainly), one in psychology or economics, one leaning toward math or philosophy, one in government "this week", one without a major. None is a freshman.

I followed my plan, with a couple of exceptions worth noting: I told a story of a little boy who was found to have severe spatial disorientation in the sense that, if he could not see something, he didn't seem to believe that it existed. For instance, when asked to lean back, he wouldn't do it for fear that there would be no chair back or maybe no floor behind him. He was in terror that he might fall into "space." He was sent to a therapist who found that he lived with a blind grandparent, and that his parents told him to return everything he ever touched to the exact samc place in which he found it. This story occasioned some discussion about how the word "space" could be defined; this discussion I will pursue further as the semester continues. Is space really a given as Kant describes, or do we collect cultural signals as to what space consists of? Can we "train in" a different "space"? What do the blind "see"?

I described the trisection problem and squaring the circle. We proved bisection of angles (they constructed the proof very easily and quickly). They then suggested that we continue on Wednesday with a discussion of Euclid, which I agreed to do....

22 September

We did do pretty much what I outlined yesterday, but there were some surprises. The first surprise was that the students knew quite a bit less about Euclid's Fifth than I thought they would. They had not seen Playfair (or maybe they could not remember it), so the idea of working toward parallelism on the sphere or in hyperbolic space seems to me now to be less possible by discussion than by actual lecturing. Also, geodesics on the sphere are still not easily visualizable to them, but that's a matter of their building intuition; clearly, I will have to spend some time in generating a few good exercises having to do with geodesics and spheres.

Another problem I encountered was in describing (or, rather, in trying to describe) curvature of a plane curve. John Smillie found a reasonable method of doing this after class when we went running, but I was surprised not to be able to convince the students of an osculating circle to a plane curve. I must admit that when I thought about describing curvature to the students before class, I decided that I would have trouble with curvature of surfaces and principal curvature, but I hadn't actually realized how difficult it would be to explain the concept of plane curvature in a simple manner without calculus.

Some points I should remember for future classes:

1. The students need some writing exercises having to do with parallelism;
2. Curvature and torsion must be discussed very slowly, with well-chosen examples;
3. Geodesics are still not obvious, and they need to be dealt with more carefully.

For tomorrow I might want to go back to parallelism (maybe à la Greenberg [Gr]), then spend the rest of the week and the weekend thinking of an appropriate set of exercises to hand out on Monday.

27 September

. . . After class Wendy and I discussed some of the content of the lecture from the mathematical viewpoint. She was especially interested in visualizing positive versus negative curvature, and why I used multiplication to determine the sign of the curvature at the given point. I explained a little bit of Morse theory, which I may want to go into in class on Monday. I also mentioned to her that the orthogonality question is one which could be asked, and how I might respond to such a question for a more mathematical audience. It is conceivable that I could ask Wendy for two projects now; one, on Morse theory, and another on the difference beween guided versus unguided discovery as an effective classroom technique, she is an education major, so this latter topic may interest her.

29 September

Class began with a discussion of the "two-dimensional bug" problem. We compared some of the students' answers, and then I went on to the question of whether we "three-dimensional bugs" can decide what kind of a universe we live in. After a while I drew a four-cube on the board, and mentioned for the first time the words vertices, edges and faces. The students were slower than I would have thought to pick up on the idea of four dimensions. They were, however, quick to get the idea that the problem may have something to do with the definition of dimension.

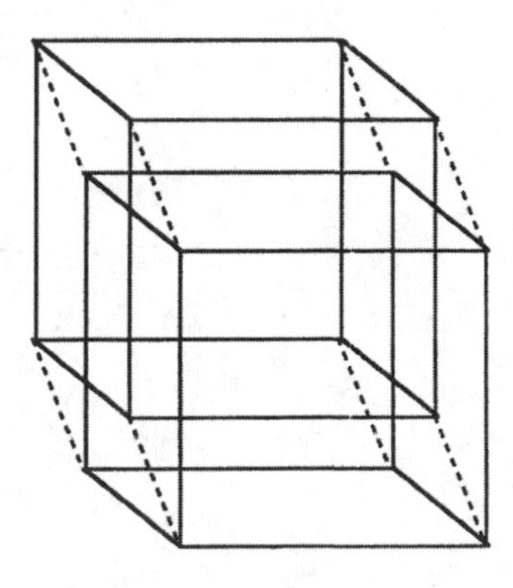

FIGURE 10

("How do mathematicians define dimension? Maybe they don't have the right idea, or maybe it's different than we think it is.") One student said "I thought time was the fourth dimension," which led us to spacetime. We tried to understand how space and time differ. This led us to philosophical questions such as whether time only goes in one direction; "Isn't time three-dimensional, because it's past, present, and future?" I drew a picture of a light cone in spacetime on the board, and mentioned simultaneity of events. I asked them whether they believed the picture accurate, but no one bit on that question.

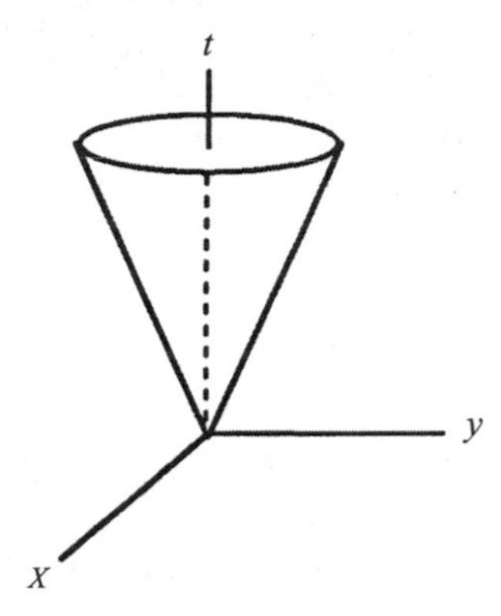

FIGURE 11

I mentioned the fundamental forces problem in physics, and how current theories lead to high dimensional spaces. Then I brought that part of the discussion back to the fact that economists and social scientists are always dealing with the geometry of multidimensions when they try to model phenomena with many parameters.

After class I joined a discussion of some students who were still wondering about perspective and vision. The question came up in class of whether we could see four dimensions if we had three eyes. This discussion continued into the question of why our eyes are in a horizontal plane, rather than vertical.

Thus endeth another day.

5

Integrating Assignments Into the Curriculum

> You've now tried a writing assignment or three. So far so good, you think, but as you hand out your most recent assignment, Cee Cee raises her hand and asks, "What do all of these writing assignments have to do with our course?"

Written assignments provide an avenue for students to develop and present their own thoughts and perspectives on the mathematics they are studying. Writing about mathematics can be a very empowering experience for students; this is certainly a primary motivation for using writing assignments and ought to be nurtured. Student work should be emphasized, not ignored during the semester. The ideas and specific answers which student essays generate should be incorporated into the presention of material in the course.

That's all fine, you say, but just how does one go about incorporating student work in a meaningful fashion? In this chapter we discuss ways to make written assignments an integral element in mathematics courses.

Consider a precalculus course, which, when taught at the university level, usually attracts students who are not particularly comfortable with mathematics. Often it is the formalism with which students have difficulty, and one place this will manifest itself is in the discussion of functions. To counteract this, give your class the following short exercise:

> Write down a definition of a function which you have encountered in the real world. Don't use functions like $f(x) = x^3$; rather, describe functions like "the temperature at time t," or better yet, "the price of a six pack of Mickey's at time t," or "the calorie content of various food items in the food court."

When the assignment comes due, ask for volunteers to read out their functions and briefly discuss each definition which is offered. Point out which definitions are clear and/or creative; when appropriate, ask the class to figure out how they might "graph" a particular function. Conclude the discussion by asking them to write a brief homework essay clearly defining their function and describing "what it looks like"; that is, what are some of its special/interesting properties. After the discussion your class will have a good idea of what kinds of functions you are hoping they will create.

In and of itself this is a good project for a class to undertake. The writing and discussion assignment will help the students understand certain aspects of functions. But don't stop there;

you can increase the effectiveness of this assignment by referring to the students' functions during the course of the semester. Use student functions on homework, quizzes and exams. For instance, after continuity has been defined, ask if "Jane's function" is continuous, or, toward the end of the course, if "Carlton's function" is differentiable. For something like the calorie-content-at-the-food-court function, there will be some serious soul searching as to what "continuous" really means.

The incorporation of tangible functions that your students create makes abstract concepts, such as continuity, more concrete. More importantly, by using their examples in class, on exams, and in homework, you directly inform your students that their work is valuable, of interest, and an important component of the course.

At times you may want to alter your syllabus in response to student essays. Often students will raise issues which you had not planned to cover, but because they have expressed interest, you can divert your course slightly in order to pick up that topic. When designing laboratory experiences, you might require a reasonable portion of written work as part of the student's solution. This helps the student understand that you value the ideas being explored in the laboratory more than the computer skills required to produce specific answers. Incorporating a written component seems to improve greatly the effectiveness of a calculus laboratory experience. (See Meier [Me 2] for a more extended discussion, and look at the coda to this chapter for an example laboratory assignment.)

In order to partially direct the students' exploration, each laboratory assignment contains numerous open-ended questions which each student may or may not choose to pursue. Having such questions prompts the students to explore the subject, and sometimes their explorations can be used to motivate material. For example, one computer lab focuses on approximating areas under curves by rectangles; the write-up asks the students to think of ways they might improve upon this method of approximation. Of course, many students realize (or remember) that using trapezoids might be more efficient than using rectangles, but once a student greatly surprised us when she suggested using polynomial approximations to the curve to cap particular regions. Essentially, she was suggesting Simpson's method, a topic not on the syllabus. It was a simple matter to replace the regularly scheduled laboratory for the following week by a lab introducing Simpson's method. The student's writeup from the previous week was directly quoted in this laboratory assignment. The material matched the graphics capabilities of *Mathematica*, and the fact that another student had suggested the technique helped generate interest and enthusiasm for the idea; some students even began calling this numerical technique the "Simpson/Small method" in honor of their classmate.

Once you have some experience using writing in your standard mathematics courses, you can begin to think about creating your own syllabus using writing assignments to *develop* material.

As often as possible, we use writing assignments to drive the syllabus of our geometry courses. In lower-level courses we generally try to present the material in as tangible a way as is possible. For instance, to return to the discussion of spherical geometry from the previous chapter:

We start our class meeting by asking if anyone can prove that the angle sum of a Euclidean triangle is always 180°. After a bit of head scratching and false starts, usually someone is able to present a proof to the class.

Following this discussion, now that the students have seen "for themselves" why the angle sum is 180°, we hand out some rubber playground balls, stickum, and string, and ask the class to experiment with triangles on spheres. We make clear to them that they need to find some notion of "straight" and some method of determining angles. Usually they will quickly develop a methodology for determining such things; for example, they may do this by implicitly using the idea of tangent planes to find angles. If nobody brought a protractor, they can estimate angle sums by simply cutting angles out of their tangent planes, laying the three angles side by side on their desks and checking to see if they add up to more than a straight line.

By the end of the class, the students will have gathered a lot of experimental evidence that the angle sum on a sphere is always greater than 180°. They will also begin to explore such concepts as *geodesics* and *tangent planes,* and we will have shown the students that there are other contexts besides the Euclidean plane where one can do geometry. These are central themes in an introductory geometry course which the students have introduced through their own work *starting from the very first day of class.*

So what does this have to do with writing? On the first day, the students are shown that their work is valued and important to the progress and presentation of the material. This is a first step toward creating an atmosphere which encourages quality work. We can build on this in-class experience by assigning a short essay:

> In our experiments we noticed that different triangles on the sphere have different angle sums. Is this true? Or were we just making errors in our approximations? Write on this, and try to justify your answer.
>
> *or*
>
> Come up with a definition of "straight line" which holds in both the Euclidean plane and the sphere. Clearly state your definition and explain why it works in both contexts.
>
> *or*
>
> Why do you think our "perfectly nice proof" that the angle sum of a triangle (in the Euclidean plane) is 180° doesn't work on the sphere? Carefully write out the proof, and indicate any steps that seem suspect on the sphere, giving examples to show why you don't think the step holds on the sphere.

We begin the next class by discussing our students' answers, and we continue to motivate and develop material throughout the semester with the students' written work.

Whether we are designing a course with the intention of having student essays drive the syllabus or simply adding short essays into a more traditional course, we should not ask our students to go through the struggle of expressing themselves only to have their work ignored in the classroom. By using frequent writing assignments in homework, in laboratories, on quizzes or exams, we will be opening a discussion in our class which will help us understand our students, and which may also help motivate our students to learn the material.

Exercises/Discussion Questions

1. Choose one central concept that you will develop in a class you will be teaching next semester. For instance, you might want to consider "continuity" in calculus. Design a group writing project which would explore that concept through a specific example. How will you implement that project?

2. Charles, the class gadfly, asks why you would ever want to factor a polynomial. How do you respond? How do you put together an assignment that *shows* Charles, rather than just telling him, why factoring is useful? How do you continue to develop this theme as the course goes on?

3. Design a short writing project for a multivariable calculus class which will have your students create their own functions of several variables. At what points in the semester might you want to refer back to these functions?

4. Let's assume you have your students write reviews of the latest section covered in the text. How could you integrate what your students write into the class?

Coda D — A Computer Assignment

Many mathematics departments are using computer laboratory assignments to enhance their calculus courses in the belief that these experiences will make their students more active learners. It seems to be true that a well-designed computer assignment allows a student to explore topics, discover connections, and to feel a sense of ownership for some part of the calculus. Throughout this text we argue that writing assignments have much the same effect. There are some dangers involved in computer assignments, many of which stem from the student using the computer to generate answers without real understanding. By adding a writing component to a laboratory experience, one can minimize such difficulties.

Below is one example of a computer laboratory assignment requiring an essay solution. This assignment is an early exercise in multivariable calculus helping the students visualize the graphs of functions of two variables and improve their intuition about the geometry of such surfaces. Throughout the assignment there are open-ended questions which the student may (or may not) choose to explore. The only requirement is that they hand in a polished written report which analyzes some aspect of the laboratory material in some depth.

The Strange World of Mins & Maxs
for Functions of Two Variables

Pre-Lab Assignment: Before coming to lab, you should find an explicit polynomial whose graph has two maxima and one minimum (like the graph in Part 3 of this lab assignment).

Part 1: We're going to examine some of the differences between maxima and minima in functions of two variables compared to what you know about maxima and minima of functions of one variable. Although we haven't defined these terms in lecture, the idea of a maximum or a minimum should be fairly intuitive.

Definition. A function $f(x, y)$ has a *local maximum* at (a, b) if $f(a, b) \geq f(x, y)$ for all points (x, y) near (a, b). It has a *local minimum* at (a, b) if $f(a, b) \leq f(x, y)$ for all points (x, y) near (a, b).

Probably a good number of you would like to believe that a function has a maximum or a minimum at "points where the derivative is zero." If so, you should probably rediscover why this "definition" isn't completely accurate even in the one variable case. You should also probably refine your statement to ∂_x and ∂_y are zero. But most importantly, you should consider the following graph:

Plot3D[x^2 − y^2, {x, −2, 2}, {y, −2, 2}]

Here both partial derivatives are zero at the origin, but clearly there is no extreme value there. Such a point is called a *saddle point*. Notice how, in the example above, the xz-trace through the origin is an up-parabola while the yz-trace is a down-parabola.

To practice spotting maxima, minima, and saddle points, you should examine the graphs of $f(x, y) = x^2 + 0.2y^2$, $f(x, y) = 0.5(x^3 - y^2)$ and $f(x, y) = (4 - x^2 - y^2)^2$. You will need to view these graphs with a variety of domains and viewpoints in order to get a somewhat accurate picture of what they look like.

Once you've played with the examples above, you should consider one more interesting function: Let $f(x, y) = 0.5x^2 + 0.5y^2 - 2xy$. The xz-trace is $f = 0.5x^2$, and the yz-trace is $f = 0.5y^2$. Because both of these are up-parabolas, one would probably expect to find a minimum at the origin, but there isn't. Can you figure out a simple, clear, and convincing explanation as to why this isn't right? (Hint: there is another trace you should consider.) Can you construct similar examples?

Part 2: Let's look at one more example. Take a nice single variable function like $f(x) = x^3 - 3x$ as your starting point. You should plot this to see what it looks like, and notice that it has a maximum at $x = -1$ and a minimum at $x = 1$. Turn this into a function of two variables by adding a y^2: $f(x, y) = x^3 - 3x + y^2$. Does this have a maximum at $(-1, 0)$? a minimum at $(1, 0)$? How has the addition of y^2 affected the maxima and minima? What would happen if you subtracted y^2 instead of adding it?

Part 3: It's time for you to start creating your own examples, instead of simply looking at the ones I have concocted. It follows from the mean value theorem that *between any two maxima of a twice differentiable function, there is a minimum.* Below is one example. See if you can't prove this statement on your own. If you get stuck, I'll give you a hint.

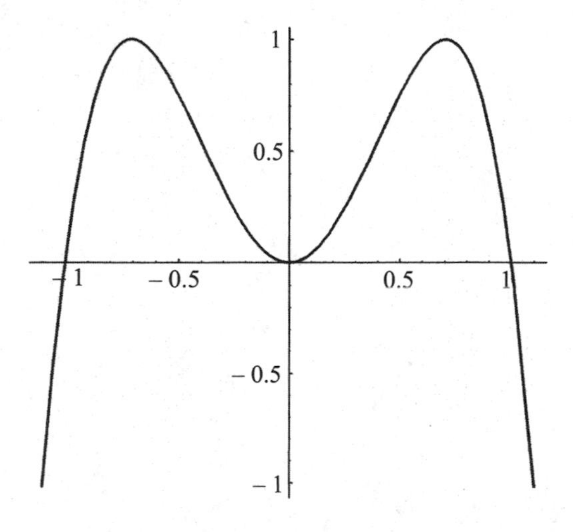

FIGURE 12

This statement is not true for functions of two variables. There are functions of two variables with two maxima and no minima. For the last part of this lab it is your job to find examples of such functions. Once you've done that, try to make a function with two minima and no maxima, or three maxima and still no minima. Do you think it would be possible to define a function where there were two maxima and no minima or saddle points?

The Write-Up: As usual, your write-up should consist of approximately two to three pages of interesting mathematics, presented in a clear and understandable fashion. It should definitely address some of what you have discovered about extrema for functions of two variables. In particular, it must contain a detailed explanation of why the function you created in part three has two maxima and no minima. High marks are awarded only if you go beyond this minimum requirement, and develop some topic from this laboratory in greater detail.

6

Grading Essays

> You're winding up your guest lecture at a conference on mathematics reform. You've been brilliant, even if you must say so yourself, at making the case for writing assignments. Suddenly, a voice pipes up from the back row, "That's fine for you, but I could never use writing assignments in my math course—I wouldn't know how to grade them . . . ."

Let's be honest; grading essays is more time consuming than grading calculus exams. But what we learn about the students and about our teaching is well worth the effort. Meanwhile, what our students learn about the complexity of critical thought, about the ways in which mathematics is accomplished and about the uses of imagination is all unreproducible with standard exams.

Grading written work will usually seem more subjective than many mathematicians are comfortable with; such grading will raise issues, such as those about grammar, that most mathematicians don't feel qualified to evaluate. Although these facts may be daunting, and they are common excuses for not assigning written work, after a relatively short amount of time you will probably find as we did that grading is no longer a primary concern.

Our students often surprise us with higher quality work than we anticipated. In large part, the key to getting such good work comes from creating good assignments; if we do our jobs well, that is, if we clearly communicate our priorities and expectations throughout the semester, the students generally write strong papers. The situation is similar with respect to grading; if we have an algorithm in mind beforehand, we can make grading easier—but still not easy.

When we grade a max/min problem, we have an internal, if not explicit, checklist of things to look for. We often start by checking the student's answer; if this is reasonable, we do a quick check through the preceding work, looking for the following points:

- Did the student set up the problem correctly, or did he just get lucky? Is there sufficient work (for example, a labeled diagram) to justify the equations they were using?
- Are the computations clear and easy to follow? (Many instructors will check that the "equals" sign is not being abused.)
- Does the student perform some final check to make sure she got a min and not a max?

We do the same kind of thing when we are grading proofs. Although the particular points of emphasis will be different from one class to the next, generally we look for an organization in which the steps of the proof are presented in a reasonable order and with correct logical structure. Once the student accomplishes those goals, we can raise the ante by starting to look for good use of notation, including clear definitions of the symbols introduced in the course of the proof and references to the text(s) of the course, and elegant and insightful, instead of merely correct, proofs.

The grading of essays, especially short essays, is quite similar to the grading of proofs and word problems. When we have a stack of papers to work through, we start grading a particular piece by scanning it quickly, checking to see if important features are present. We then check the essay more carefully, examining the parts as well as how they fit together. Items we often look for include:

- Does the student have a clear introduction which presents the main points of the essay? Is this introduction actually interesting?
- Is there a clear conclusion?
- Is the essay well organized and easy to follow? Does the student's argument support the introduction and conclusion?
- Are mathematical items (such as variables) introduced without explanation? That is, do you as reader ever say, "Wait! What's going on here?"
- Is there clear indication as to why the student is performing his or her calculations?
- Are the mathematical arguments clear and well stated? Is sufficient detail shown? Are the arguments correct?
- Is the paragraph structure supportive of the student's general discussion?
- How's the spelling, grammar, and punctuation?

Practically all of these items are things we naturally do when grading word problems or proofs. The only thing we might find "scary" is checking paragraph structure and grammar. But remember, the key idea of paragraphs, grammar, spelling, etc., is to help make a person's writing clearer so that we can better understand what the author is trying to say. Our students will take other classes aimed at improving the technical aspects of their writing. We suggest indicating where poor grammatical structure detracts from the mathematical argument and otherwise don't place undue stress on this issue.

Some instructors take the idea of a checklist quite seriously. For instance, Annalisa Crannell [Cr] hands out coversheets detailing her grading priorities with each writing assignment. When she grades a paper, she simply comments on how well the student accomplished each task on this cover sheet.

While a checklist makes priorities clear to the student-authors, it can sometimes be a bit too formulaic. Neither of us grades "standard" problems in a checklist fashion, assigning certain points for the successful accomplishment of certain steps in a problem, unless we really have to. We try instead to grade for the "sense" of the assignment. For a min/max problem, we basically decide if the student has presented a coherent and careful description of where the maximum value occurs, and if the student hasn't, how reasonable is the work presented? We tend to grade essay assignments in the same way. For example, let's say we've asked our students to write an additional example of a volume computation to be used in the next edition of their calculus text. In grading the students' examples, we would expect to find correct computations. We would be more interested in the *originality* of the example: would the addition of this example give the reader a better understanding of the topic? We would also ask how *clearly* the problem is set up and solved: could a student follow the example? Finally, to a lesser extent, we would ask ourselves how comfortable the student is with the text and topic: can the student write out an example in the rather formal mode of a calculus textbook? We would not, however, grade on a scale of four points for originality, four points for clarity, and two points for formality; rather, we would read through a particular solution, decide how well it fits inside the framework of our expectations, and then grade it based on how well it holds up on its own. If the example is not particularly creative but the presentation makes certain steps clearer than the text did, the assignment will get high marks. Similarly, a highly creative example which is not particularly clear in its presentation due to its difficulty will also get good marks as long as it has been done correctly.

Assuming we choose not to use a checklist, we still need to inform our students as to what our priorities are when grading. One way to do this is with our comments on their papers. If we want clear and well organized papers, we take off points for obfuscation; if grammar is important to us, we indicate grammatical errors and in our concluding comments try to point out the general principles the author seems to be missing. If, in our concluding comments, we praise certain aspects of a piece, such as originality and/or clarity, the student will continue to work on creating original and/or clear essays.

Just as when we grade traditional homeworks, we do need to be consistent. This applies both to different students' work within a single assignment, as well as to individuals' papers throughout the semester. The most effective way to do this is to decide, early in the semester, what we wish to emphasize in our grading of written work.

As we are designing an assignment, we ask the following questions:

- What mathematical concept am I trying to evaluate with this assignment?
- Is the assigned length appropriate?
- Are the students prepared for this assignment?
- What relevant topics do I value; which of those are important here?
- Does my assignment indicate my priorities to the student?

Once our assignment has passed this checklist, grading it will become somewhat easier. We mainly have to check if our students accomplished or superseded the goals which we set for them. Of course, the question of how to grade is interconnected with many of the other ideas we are considering; we will be returning to the topic often as this text continues.

Exercises/Discussion Questions

1. When you grade proofs, do you simply grade them based on whether or not a complete proof has been presented? Or do you take off points if certain steps are not clear, or if the student is making a proof more difficult than is necessary? What are your priorities when you grade proofs?

2. If you assigned a one-page essay worth 10 points and one essay is very creative but contains multiple spelling mistakes and some grammatical errors, how many points do you think you would give the essay? Why? How important is grammar to you?

3. Design a short checklist for your own essay grading. Which topics are most important for you? How do you assign relative point values? In what ways does your checklist address the "creativity" question?

4. Go back to our general checklist of things we look for when grading essays. How different is this checklist from one you might make for refereeing papers submitted to research journals? In what ways will your requirements for essays be similar to what you want to see in a research article? In what ways will they be different? (There should be *some* similarities and *some* differences!)

Listening to Others

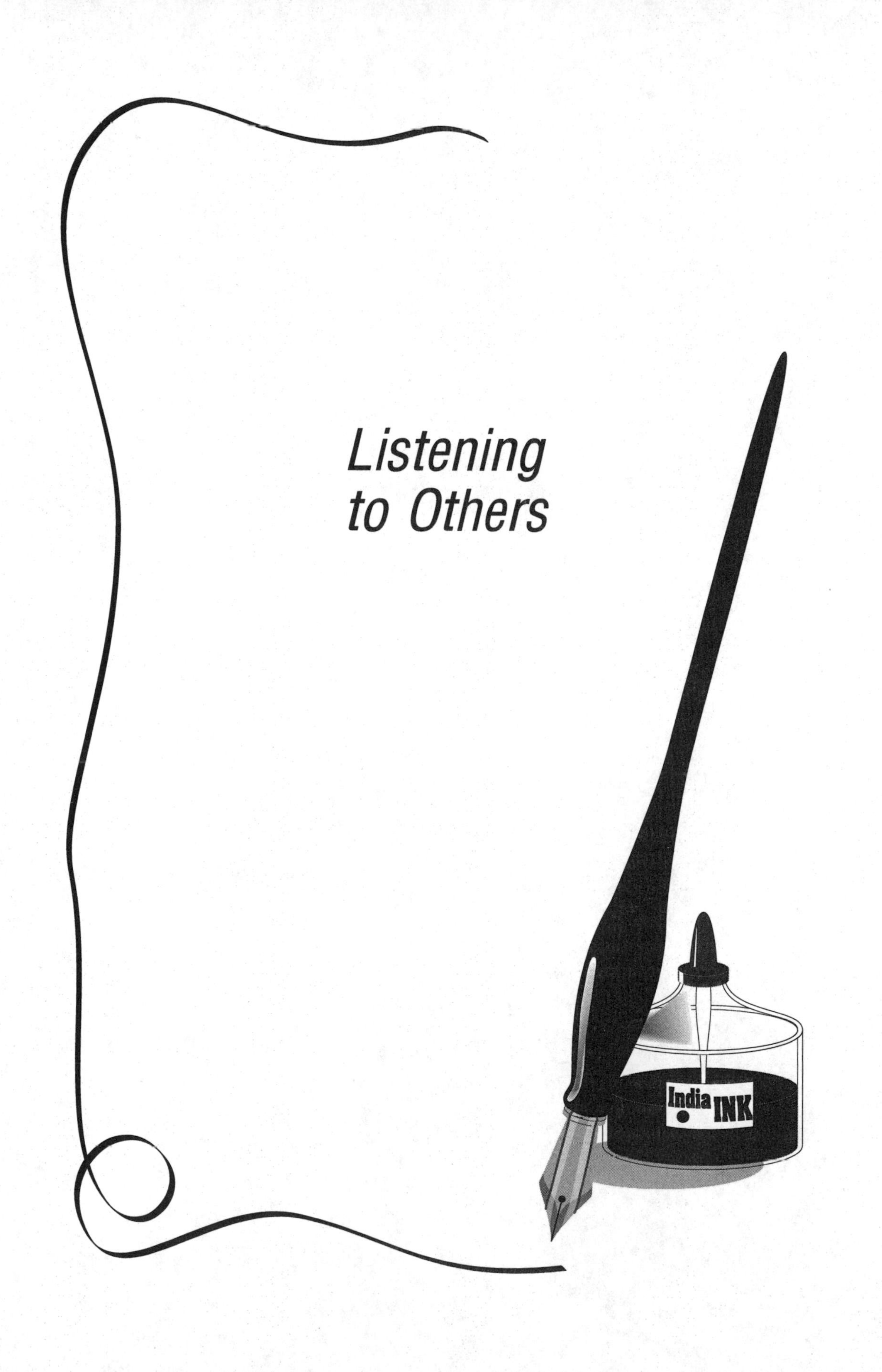

7

Learning from the Writing Faculty

You meet a colleague from the writing program at lunch. She says that she's heard you're using writing assignments in your mathematics class. Pretty soon, you find yourself being asked about "narrative form" and "reader response." You yourself don't know how to respond.

Sometimes mathematicians see their subject as being so self-contained that they think they can't get help from outside sources. This is the wrong tack to take; faculty in other disciplines have already thought through a lot of the writing assignment questions that are just beginning to bother us, and we can profit from their experience. So, we can sit in the math lounge, thinking to ourself, "How I can ever get them to stop writing this garbage. . . ?" Or, we can take a walk to the writing lounge (the coffee's probably better there anyway).

The criteria for a quality expository paper are not the same as for the standard research paper. To get some idea of what kind of results we would like to see from students, we must adopt different standards. But what kinds of issues should a mathematics instructor consider? In this chapter, we will begin to discuss such topics as audience, format, process, product, induction and deduction, showing and telling, narrative line, voice and tone, and types of papers. In later chapters we will be using this vocabulary to deepen our awareness of how writing and mathematics can be made to interact.

An expository paper should be "interesting." The question is, "interesting to whom?" And, "what is interesting?" Further, and unsurprisingly, for mathematicians, "interesting" should have to do with mathematics.

Every paper has an *audience*, even if that audience is only the instructor. However, it is helpful to get the student to think in terms of a broader group. Of course they are writing the paper as an assignment for us, but ask them to try to write for their parents or grandparents, or maybe for their reasonably intelligent, but not necessarily mathematically inclined, roommate. Get them to ask: "What interests my grandmother?" Would she get the point of the explanation they're giving? What kind of questions would she have about the exposition? How could your student make a parent interested in a discussion of, say, quadratic equations or the Fourier transform? And, if it's impossible to interest the parents, how about the biology-major roommate?

We will talk about the topic of audience in more detail in the next chapter.

To help students arouse the interest of the audience, we can suggest a very basic *format* for early papers—not that more advanced papers will fit this format, but we will give the fledgling writer some help in starting. We have found the following a useful first template:

- I'm going to tell you about [TOPIC] . . . .
- **It's important because** it can be used in the following applications, OR, because I don't quite understand it, and I want to talk it out. . . . Here's the [TOPIC]. . . .
- Let me give a real-world example, to give you an idea. . . .
- Here's where it occurs specifically, OR, here's where the example is helping me try to get a handle on this difficult topic I still don't necessarily understand.

There is a famous type of paper which has been known and discussed since antiquity; it's the *five-paragraph essay*. Many students (and, no doubt, many of us) will have been taught this form of writing in, say, high school. The basic form is as follows:

Paragraph 1: Here's my claim (the thesis statement).
Paragraph 2: Here's my supporting evidence.
Paragraph 3: Here's what my detractors claim.
Paragraph 4: Here's why my opponents are wrong.
Paragraph 5: See? I've proved my case.

One of the authors (Tom) went to a writing faculty conference a few years ago at which all the faculty sat around a large conference table and spent a half-hour roundly criticizing the five-paragraph essay, as well as any poor, benighted individual who might teach it. Let us take that back—all *but one* of the faculty criticized this form of writing—Tom did not. Oh, he didn't defend the five-paragraph essay, partly because he did want to leave the room alive, and partly because he doesn't believe that every piece of writing can fit that mold. But, on the other hand, he does believe that there can be good reason to let students use some externally derived form, if only so that they can work away from it—so that once they have used it, they can see how another form can be more appropriate to the topic they have been given. Yet, there is something to be said for letting students start from the ancient, tried, and (somewhat) true safety net of a predetermined format for papers—later on, we can tell them that there are other ideas, other forms, other ways of describing the world out there.

Writing faculty often talk about how students must go through a *prewriting* phase before they actually begin an assignment. This prewriting doesn't consist only of bad first drafts; it can also include *classroom discussion, student talks* and *conferences.* In fact, one of our favorite assignments is to send students out to the quad to simply **think** about the topic or discussion just assigned, thus enforcing the idea that they need not immediately start putting pencil to paper (or maybe "putting fingers to word processor" is a more apt phrase nowadays). Since we find prewriting to be very important, we will be returning to this topic again in later chapters.

One of the more venerable topics of debate among writing faculty is that of writing as *process* versus *product.* Once upon a time, college was for the elite, and writing was a skill which was supposed to be known *before* students attended freshman orientation. Writing faculty were the part-timers hired by the English departments to teach night school and the irremediable. In this universe, freshman writing was about the "final product," about Shakespeare and Maugham, and the Aristotelian model five-paragraph essay was the only format. Our goal here is not to ask whether such a utopia ever existed; needless to say, it no longer does today. At some point, college faculty decided, because of a number of motivating forces, that the goal-oriented *product approach* ("Write a five-hundred-word essay on whether Hamlet was really mad or not. . . .") was not achieving results for the then-current college student body. Soon, writing courses became much more interdisciplinary (there is maybe no better example of this than a text on using writing in mathematics), designed to reach a clientele looking for "relevance." Partly as a result of this, writing became more formless, more of a "search for understanding" of diverse topics than a deep delving into literature texts. The writing assignment became more of a process through which the student could come to an understanding of what he or she saw, felt, intuited, on any particular day, in any given assignment. For some instructors, "process" became the *entire* goal—the argument being that every text is incomplete.

Here, we take the position that writing *is* a process—write a short draft, revise, discuss, revise, question, write more, look for sources, start again—which leads to a product: the final paper. And then, that "final" paper leads to more thought, i.e., process, maybe after class has ended. And that can lead to more product, and it goes on, and on, and on. . . . (For more detail, see Elbow [El 1]).

In Chapter 14, we will return to this debate over process versus product; for now, our goal is simply to be aware of the discussion, so that we can think about how it fits into our own global sense of *your* assignments and courses, which is surely the most relevant part of the debate—for us.

Another useful concept for getting the student accustomed to writing expository prose is that of *displacement* or *detachment.* When other people explain things to our students, what do the students like to hear? Do they prefer simple examples of new topics to general statements? Does the showing of relevance to one's life attract their attention? So, what should that tell students (not to mention *us*) about how they should write?

It took us a long time to get used to the idea that when writing faculty talk about *induction* and *deduction* they don't mean exactly the same as mathematics faculty do when they talk about the same topics. For writers, induction means "from specific to general," while deduction is just the opposite.

Most college students are in love with deductive writing, while faculty tend to prefer inductive. Faculty hope to teach *critical thinking* derived from informed opinion and valid evidence. The student's reasoning seems to be: "Now that I'm in college, I'll be wanting to show my erudition; the instructor will want me to make general statements which I can then back up." Meanwhile, the instructor is looking for "vivid prose, well-chosen examples, writing from experience." So, the student writes, "Everyone has a name, and every name has a meaning. Take my name, for instance. . . ." Meanwhile, the instructor is looking for "Call me Ishmael."

So how do we get our students to write clear, vivid prose? We might ask them to *show, not tell.* For example, look at the paragraph on "format" above. Notice that the statement "It's important because..." is in boldface. This is because, if our student writes, for instance, "I want to tell you about Fourier series because it's an important topic," our student's audience will say either, "Maybe for you, because you have to write about it; but not for me," or, "Well, if it's so important, maybe you'd like to tell me why!" A better sentence is: "Fourier series is important for applications to heat transfer, for instance." But an even more interesting beginning is: "Have you ever driven by those swimming pool lookalikes that are really pollution control devices? Did you realize that they have something to do with Fourier series?" Now the reader is being shown that he should have some interest in Fourier series, and if he doesn't want to know about the topic, the fault is unlikely to be that of the writer.

Another mistake that authors often make is in claiming that too many ideas, proofs, and the like are "obvious." In these instances the writers are *telling* when they should be *showing*. If writers do not want to show why all isosceles triangles have a pair of equal angles, then they can put the details in an appendix or leave them for an exercise. They can even cite a reference. Generally speaking, however, they owe the reader as much explanation as possible.

All students would profit from considering *narrative line* as a story. After all, they are not just proving theorems, they are telling a story about some aspect of mathematics; and that story has a beginning, middle, and end. They must find the best way through that narrative—one which keeps the readers' attention. They need to keep asking themselves: "What do I need to get the story from point 'a' to point 'z'? What will I need to make clear to the reader to get to point 'z'?"

A lot of mathematics, even at the most advanced levels, begins with a narrative, a "story line" we give ourselves and our colleagues about "what has to be true" and why. We then refine, make precise, answer objections, offer examples, until our work begins to approximate Proposition 49 of Euclid's *Elements*; that is, until we forget that we began with a narrative.

If we working mathematicians begin with narrative, how much more is that true for our students, who have not yet learned the rigors of a mathematical argument. Thus it is very logical for us to ask them to learn mathematics through the act of writing their experience in narrative form.

We will return to the topic of narrative in the last chapter of the book.

Tone is yet another concept. "This proof is really cool" is, like, maybe an O.K. statement to do in a group discussion, but, you know, unless humor is intended, it doesn't really make it so good in print, "if ya get my drift." Similarly, a statement like, "The following proof is a sophistical analysis of the intellectually invigorating conceptualization of exterior spatialization," may also say that the proof is "cool," but the hypersophisticated, recondite style of discourse is going to sound stilted and unlikely to invigorate an audience. To help students find the right tone for their writing, we try the idea of having a first assignment which asks them to write, seemingly "off the cuff, in their own words." Example: "Write me a paragraph or two about what you might like to see in this course." In this way, we get more natural language, and we are signaling to the students that they can write *to us from their own perspective*. Of course, we are more likely then to have the students err on the side of the informal, "this is cool" form of discourse, but that's a much easier form to deal with than the opposite. In fact, we appreciate such language, especially in early papers; students who are willing to write like that to the instructor are both relaxed and interested in their papers, and capturing that excitement is what such an assignment is all about.

Along with tone, any writing instructor will tell you, comes *voice*. This is harder to describe than tone, but we will try.

Did you ever sit in a lecture or read an editorial, and when the speaker or editor says something you disagree with, something inside you says, "Fat chance?" Conversely, did you say to yourself, "I know that from personal experience, because. . . ?" In either case, that little voice was *your voice,* speaking to you. We want to encourage that voice in our students, within reason, because it's the "real you" coming out into public view. And, what we want to encourage from our students is that those who have small, little voices buried deep back in their brains will allow those voices to grow inside those brains enough that maybe they'll even equal the size of their brains by the end of the course.

Of course, we have encountered one or two students whose voices are much larger than their, not to mention our, brains. People who are larger than life need to learn how to question themselves, need to be challenged in meaningful ways. These students can be useful in classroom discussion, of course: "Does everyone agree with the student who said that geometry has never been useful in any way?" Generally, however, most students will need encouragement, not discouragement, and will appreciate whatever help we can give them in finding their way of approaching mathematics, writing and learning, that is, in finding their voice.

Entire volumes have been written on the topics of voice and tone; if we want to start a real debate in the English department coffee room, we just ask any two faculty members to define them and describe how they differ. For us, before we get into trouble, we will tiptoe out of the coffee room and into another topic . . .

The papers you can assign fall into a number of categories. Three useful types, if over-simplified, to remember are the *personal statement,* the *informational essay,* and the *argument paper.*

Personal statements, which are often the easiest to write, can be of the form, "I like geometry because. . . ," or, "My favorite theorem is. . . ." These papers can be simple for the student because they require no library time, no external sources, and because the student isn't required to "prove" anything—thus, personal statements form a good starting point for writing in a course.

Informational essays are only slightly more difficult. The students collect some facts at the library or off the internet about, say, the golden ratio or the life of Bernoulli, and they write them down in a somewhat coherent manner. The collection process teaches them about sources and materials, and provides good preparation for the third type of paper.

Once the students are used to collecting facts and telling you whether they like them or not, it is time for them to try an argument paper—one where they are asked to analyze and synthesize the data collected. For instance, continuing with the theme of the golden ratio, one such paper might be: "Objects found in nature which follow the golden ratio are truly the most beautiful forms which exist." A thesis statement for a paper taking the opposite side of this same question might be: "The golden ratio is an artificial construct which only derives its status from the fact that so many people throughout history have assumed it has something to do with esthetic theory." (There is a sense in which every mathematical proof constitutes an "argument paper.")

Argument papers are the most difficult to write; they ask the students to collect information, decide on a meaningful question about that information, and then take a position on that question. They must then analyze the data, synthesize ("boil down") other people's arguments, and then make a convincing case for their thesis. In so doing, the students are compelled to learn a great deal about the subject; after all, informing, analyzing, and synthesizing are three complex modes of thought. Therefore, an argument paper is a significant tool for aiding understanding of any subject, including a mathematical one.

Having made a somewhat coherent argument for argument papers, we want to admit that the three categories above are not really mutually exclusive. There are plenty of personal papers which contain argument; in fact, we would claim that any *good* personal paper must have a convincing thesis at its center. Further, solid arguments start with concrete factual information. We have found in our own teaching that explaining the distinction among the personal, informational, and argumentative categories affords students the opportunity to write deeper papers with sharper theses than we would otherwise obtain.

We will return to the problem of how to construct meaningful papers in Chapter 10 when we discuss questions which test students' cognitive comprehension.

Exercises/Discussion Questions

1. Design your own template for a basic student paper. How does it differ from the one above? To what extent does your template differ because the course you are teaching is different from that "described" above?

2. *Reader response* is another term often used among writing faculty. What do you think it might mean, and how could it pertain to the students you see?

3. Talk to a faculty member in your writing program. Ask that faculty member what are the "buzz words" among the writing faculty. What can you learn from these; how will they help your students?

4. Basically, this entire chapter could be summarized in one succinct paragraph, something about how you can help the students write "more interesting papers by using techniques learned from writing instructors." Write such a paragraph.

What additional themes might we have left out?

8

Knowing the Audience

He must understand the nature of the soul, must find out the class of speech adapted to each nature, and must arrange and adorn his discourse accordingly.

Plato, *Phaedrus*

In the previous chapter, we introduced the concept of audience. Now let's expand on that discussion.

When designing effective writing assignments, consider the audience. We mean this in two ways:

(i) Think about who the students are and what they should learn;
(ii) Let the students know for whom they should be writing.

These two ideas are closely intertwined. For example, compare a real analysis class and a first-semester calculus class. At approximately the same point in the semester, we try to convey to both classes what a derivative is. Both in calculus and in real analysis we have asked the following question:

Compare the standard derivative $f'(x)$ with the *ampersand derivative* given by the formula:

$$f^{\&}(x) = \lim_{h \to 0} \frac{f(x+h) - f(x-h)}{2h}.$$

Are these the same for every value of x and for every function f? What's different about $f'(x)$ and $f^{\&}(x)$?

To ask this one question of both classes is a mistake. It is a good exercise for a real analysis class; but even if this problem is toned down for the calculus course by concentrating on the examples $f(x) = x^n$ and $f(x) = |x|$, it simply does not work for first-year students. Thus we have learned the hard way that we cannot use the same assignment for two different audiences.

The fundamental focus of calculus classes and real analysis classes is different. Although calculus is in transition, it still seems to focus on general concepts and computational ability. Real analysis, on the other hand, emphasizes the careful use of precise definitions and theorems to construct proofs. Probably no matter what sort of selling job we might use on the ampersand derivative problem, it just won't be appropriate for a calculus course. This problem is designed to force its solver to look hard at the definition of a derivative and the algebraic and geometric concepts behind it. It is not a problem which will aid the average first-year student in understanding derivatives, and it certainly won't help that student compute derivatives.

The bottom line is: What we hope the students will achieve, and what we believe they are capable of, is different in different courses. It follows that the nature of the writing assignments you use should be different also.

In most service courses we want students to acquire the ability to discern how and when to use certain computational techniques. (We realize that is a somewhat optimistic statement.) Most people don't like simply memorizing and applying rules, although they may prefer this to independent thinking. In such courses short narrative projects are quite helpful, because we want to bridge the gap between the students' views of the universe and the computations we are now asking them to perform.

For instance, in a business calculus class we are about to enter the section on antidifferentiation and definite integration. This is certainly going to be a tough audience for selling the idea of approximating area under a curve by rectangular approximation, and we may or may not have ever mentioned limits when we discussed differential calculus. This could be a place for us to be efficient, to side-step the issue of "what do you mean by area," and get to the next part of the syllabus.

But what if we can't do that? We might want to look around for a short writing assignment that will get the class thinking about area approximation. For instance, we noticed one night at dinner that one of the dining areas at Lafayette College is shaped roughly like a semicircle surmounted on a rectangle; the floor plan is made more complex because certain areas are at different levels than others, and there are stairs connecting each level. So we asked our class to go to that area and approximate the total amount of carpet they would need to cover the floor, the stairs, and the sides of the stairs. Most students did a fine job of realizing they needed to "chop up" the floor plan into familiar pieces and then add up the areas of these pieces. A few students even worried about arc-length issues when they approximated the cylindrical shapes formed by the edges of the stairs.

Such an assignment made our lecture on area under a curve go quite smoothly; the standard method of approximation by rectangles seemed natural to the class, and they actually were somewhat interested in the "primacy of rectangles." "Couldn't you use semicircular pieces, or triangles like we did?" This is certainly not the sort of question one expects from a business calculus class.

Even within such a service course, assignments can vary according to the particular students. One assignment that would work for a typical group of students at a small college might be of little interest, or require too much background material, for a group of students at a public city university.

Another instance:

Lafayette is a small college which plays Division I sports. About one-quarter of the student population is on a varsity team, and most of the rest are playing intramurals. Thus at Lafayette incorporating sports into exercises can be effective. For example, John had a senior varsity basketball player, Sharon, in his finite math/precalculus class. One day John asked her what her free-throw percentages were for her first three years on the team. The class was then divided into groups of four, each group being assigned the task of writing a letter to the varsity women's basketball coach. The letter began by explaining that they were going to model Sharon's free-throw percentage by a logistic curve, and then carefully present their computations and their estimate of what her free-throw percentage ought to be for the upcoming season. Because the class was interested in basketball and had already done a number of short writing assignments, they undertook the assignment with little need for explanation. (This example might not work at a college with no Division I sports program and limited campus enthusiasm for athletics.)

We need to have a clear vision of who the students are before deciding on a particular assignment. Conversely, once we have decided who they are and what we want them to contemplate, there is the second question of audience: Whom should the students be writing for? This really could be rephrased as: What aspect of the material are you trying to emphasize?

Often, especially with shorter assignments in first- and second-year courses, we encourage students to write informally, if not "creatively." Most students struggle with basic ideas and with the abstraction and formalism presented in a typical classroom. Often they become more independent learners if they are allowed to create and explore their own metaphors for the concepts being discussed and to apply them to the outside world. Thus, informal writing assignments in mathematics courses can help students become familiar and comfortable with the ideas and tools being presented, without forcing them to worry about issues of format, grammar, and style.

All students are aware that they are writing their essay for their professor. The sort of highly edited and lifeless academic prose that results from this will not increase their comfort with the mathematics being discussed. Many of the initial essays in calculus classes start by saying something like: "The purpose of this assignment was to find the slope of tangent lines to various curves. The tools used were paper, pencil, and *Mathematica*. One begins by looking at secant lines . . . ." There is little evidence here that the student has absorbed the notion of tangent lines or that the student is at all enthusiastic about the topic.

In order to avoid such writing it is often useful to assign a particular audience for each paper. For example, in a multivariate calculus course students work on a computer-integrated project based on three-dimensional geometric linear programming. The assignment asks them to find the most cost effective combination of methods to clean up a chemical spill given relevant parameters for three available methods. Their instructions for the write-up are:

> Your write-up should be an explanation to your boss as to why you picked a certain combination of methods for this project. You should keep in mind the following guidelines:
>
> 1) Suppose you are asked to justify your methodology to the Senate. If some senator claims you did a poor job, this write-up is your justification of why you did what you did.
>
> 2) Not everyone reading this report will be technically trained. Do not assume they already know and understand geometric linear programming or that they are interested in the *Mathematica* commands you used in solving this problem. They will, however, be very interested in knowing the general method you used and why it works.
>
> 3) Your report should contain a clear-cut *answer*. That is, you need to clearly state exactly how many cubic meters of waste will be processed by each method.

The assignment is intended to be entertaining, but also focused and clear enough that the students can engage their creativity while not straying too far from the "real" assignment, which is to explain the idea behind geometric linear programming.

Many of the write-ups are in letter format—usually as an internal memo to their boss ("Dear Boss ..." is a common first line)—but in a couple of cases, as letters to the honorable senators of the student's home state ("Dear Senator Kennedy, The following report is respectfully submitted in response to your accusations concerning my most recent work dealing with the chemical spill in Easton, PA... ."). Notice that the students are certainly exploring their own voice, in the context of a linear programming assignment!

The students' choice of the intended reader can be crucial in generating good writing and improving understanding. Requesting students to write explanations to friends is usually a good choice. "What examples would *you* choose to use to explain an idea?" "Can you relate this to something you are studying in other courses?"

Bob Montgomery [Mo] uses a magic genie to help break the ice on the first writing assignment. For instance, here's one he uses in differential equations.

> On Thursday night, after you've finished your homework and are back home relaxing, you meet up with the Pardee Genie, $\delta\epsilon$. You have a long and interesting discussion about differential equations. During the evening $\delta\epsilon$ says the following things:
>
> > *"Differential equations are really easy. They are all basically the same—all you really have to do is put all the y's on one side of the equation and all the x's on the other side and then integrate."* and
> >
> > *"Oh, that's right, there is that really cute differential equation modeling weights attached to springs. Of course, that says that the weight always bounces forever, and we know from experience that that never really happens."*
>
> Write a short response to *one* of $\delta\epsilon$'s statements. You should use specific examples to help illustrate your point.

In response to this assignment some students hand in standard academic essays, which are fine, and they receive full marks; similarly, on a more formal assignment others may be fairly creative and informal, and they too might receive full marks. For the Pardee Genie assignment, one student, dubbing himself "the great playwright," presented a complete dialogue between himself and $\delta\epsilon$. The character *Me* in the text below is the student-author.

> Me: Look at this D.E.: $y" + 0.5y' + y = 0$. That would be a really difficult equation to solve by separating x's and y's. You'd have to make an auxiliary equation for this one, and even so it'd be tricky because you get imaginary numbers.
>
> > ($\delta\epsilon$, *feeling sheepish, conjures up a teddy bear to make himself feel better.*)
> >
> > ...

Making the writing informal deflects attention away from the instructor by explicitly asking students to write for a less threatening audience. At the same time, we should be aware that the student may stray too far afield from the original question and may try to avoid the mathematics that they should discuss by *overemphasizing* the creative writing aspects. For instance, a student once presented his work in the format of a *New York Times* article; although the idea was original, too much energy had gone into the format and not enough into the mathematics. Generally such difficulties can be easily corrected by the final comments and grade.

By asking our students to address their essays to friends, genies, or senators, we are indicating what is an appropriate level for their paper. Generally we try to pick an audience that will require the level of detail and explanation we want our students to give. If we want details, we have them write a technical report to their employer; if we want the basic ideas involved, we ask them to write to their parents; if we want a good explanation of a process, perhaps writing to a mythical friend taking the same course at another college would be most appropriate.

Being clear about audience helps students focus on important aspects of the assignment. This is true not only for informal writing assignments, but even for writing proofs. Often students will become lazy in this kind of writing, assuming that the instructor already knows all the material and doesn't need to see a carefully written explanation. Having students write their proofs for their classmates or other mathematics majors can improve their presentation.

Of course, proof writing is not the only form relevant in upper level courses. In order to add breadth to a course and to develop our students' ability to read mathematics, we sometimes have them read and then write on articles in the *American Mathematical Monthly.* Unless we have truly exceptional students, we will have to create some writing assignment which will force them to do the reading. For example, we occasionally have each student explain in a paper the ideas and techniques by using an example which does not occur in the article. The target audience can be the other students in the course; this choice of audience will help the students to break their articles down into language they are comfortable with. Then they can present the material at a level appropriate to their abilities.

Our audience includes our class, and it is important to consider their interests, background and motivation when creating writing assignments. Establishing a well defined audience appropriate to the assignment helps create clearer, more focused and energetic prose; then, by concentrating on writing to explain, as opposed to writing to impress, the students create more coherent mathematical work.

Exercises/Discussion Questions

1. Think of your service courses where the students are generally enrolled in order to satisfy some distribution requirement. How can you discover your students' outside interests? How can you design a simple assignment which they will be willing to spend time working on?

2. At the time of this writing there exists a great deal of interest among the business populace in the "derivatives market" because the government is considering regulation of this part of the financial industry. How can you find out about this market, and what its connection is to derivatives in calculus? Is there some good source material, such as a *Wall Street Journal* article, which a business calculus class would find interesting to read and respond to?

3. Your linear algebra class is resisting the idea of the space of polynomials as a vector space. Can you think of a writing assignment which you could include as part of your weekly homework assignment which would get the students focusing on the idea of polynomials as vectors? Whom should the students write such an essay for?

9

Prewriting and Writing

Students have written one or two assigned papers for you now. The papers, however, appear "canned," forced. You want to engage them on a different level, break them from the mold of copying from another book in a hollow voice. What do you do?

"Prewriting" is what we write when we're getting ready to "really write"—whatever "real writing" is. Prewriting consists of our notes, musings, outlines; our "thoughts to ourselves." It's all the **cogitating** we put into a paper before we begin that draft-we-wish-was-the-last. Often, it includes that draft, too. In short, prewriting is the thinking that's necessary before the idea and the paper become ours.

One of our favorite assignments is to tell students to "walk around the quad three times and just think about" some topic; that's prewriting. Conversely, we ourselves sometimes go out on a run to think about how to shape a letter of recommendation; that, too, is prewriting.

The process continues forever; someday, when we are "old folk," we hope to know so much about our favorite topics, we won't even notice that we've spent 30 years in the prewriting phase. For now, while we're still young, we need some prewriting.

Now, if *we* need prewriting, just think about how much our first-year students need it. So here's how to get them to go for it:

Let's assume we spent two days in calculus lecturing on the fundamental theorem. To get our students thinking constructively about that theorem, we start the third class by handing them a single sheet of paper with the following question attached: "What's so 'fundamental' about the fundamental theorem of integral calculus?" Now, of course, we have given some hints as to some possible answers to the question in our lectures, but we also know that this question is formless enough that there isn't any single, correct, universal answer.

We read the question to the students. After letting them have some time for free discussion, we ask them to formulate a written response on the remainder of the single sheet of paper we've given them. We tell them that we just want them to write what's in their heads, that there's no "right answer," that we'll collect their responses, but "won't grade them too seriously." We say something like "I just want to see what's been coming across from the last couple of lectures." This last is literally true, and at the same time it re-allocates some responsibility: *from* the student, for having "the answer"; *to* us, for our lecture.

We take the student responses back to our office, or, better yet, to the coffee room, and read them to ourselves—don't *grade,* just read. We circle good statements, writing a short comment next to a particularly intuitive word or phrase, add a question mark where things get fuzzy or obviously wrong. We needn't make stentorian comments about "wrong" ideas; we can deal with them later. We give people minimal grades—maybe "check, check-plus, and check-minus." If someone did absolutely nothing, we don't give that person any grade. Later in the semester, grades may become important; for now, getting students to express themselves is foremost.

Then we start our next class by discussing the papers for 5 or 10 minutes, bringing up some of the ideas we noticed, the intuitive leaps, questionable comments. We are fuzzy, but not misleading, about who had incorrect, unintuitive, counter-intuitive concepts (After all, who put them into those young, impressionable minds? But then, as a student once said, "You shouldn't be so hard on yourself, professor.") We answer further questions the students might have about the assignment. When the "smart one" asks how we would have answered our own question, we have an answer ready. (Well, we read all theirs, right? So, we should have seen a lot of possibilities.)

Let's take stock of what just occurred:

- By asking the students to "write something small," and giving them a little time to do it, we have been able to find out a large amount of information: we have an idea of what the students understand, and what is confusing them.
- Notice how the questions we ask about "intuitive leaps" and the like constitute a lesson plan. They motivate the student to discuss those topics we instructors consider important.
- Further, we're closer to knowing whether lectures are coming across. We have been able to engage the students in a dialogue, and, if we are at all lucky, we will have gotten most of them much closer to an intuitive understanding of a major concept in calculus.

Not a bad result, eh? It's even possible that some day two years from now a student will walk up to us on the quad and say, "I still remember that assignment you gave us about the fundamental theorem. . . ."

Let's see the method once more, this time concentrating on a student paper in the *From Space to Geometry* course discussed in Coda C.

From Space to Geometry—Classroom Assignment.

> Last time I talked about a triangle on a sphere. How would you describe my definition? Does it make sense to you? Why, or why not? Would you suggest an alternate definition, or do you believe that none can be constructed? And why would it matter at all whether such a thing as a triangle on a sphere could exist, anyway?

Figure 13, on the next page, is one student's response. Figure 14, on page 54, is the same response with my comments added.

Let us now try to analyze this assignment and answer.

The core of what this student is saying is that there is an abstract, objective thing called Euclidean geometry, and she is going to *push* this abstraction down onto the surface of the sphere; that will be the way she defines the triangle—starting with a Euclidean triangle, she is allowing it to project itself onto the sphere to get three curved lines and three vertices.

Further, she has used the intriguing term "volumetric"; she hasn't decided yet whether the triangle is *in* or *on* the sphere. There are two alternatives this student is still pondering; namely, whether the sphere is two-dimensional or three. She hasn't decided which one she believes. Thus, we have circled this word "volumetric" and asked her a question about it; "But aren't we doing two-dimensional stuff?" is designed to ask her to think more about the source of

Math 150—Classroom Assignment

Last time I talked about a triangle on a sphere.

How would you describe my definition? Does that definition make sense to you? Why, or why not? Would you suggest an alternate definition, or do you believe none can be constructed? And why would it matter at all whether such a thing as a triangle on a sphere could exist, anyway?

Assuming Euclidean depiction is objective

that which supperimposed onto a flat surface from the spherical surface, would form a triangle on the flat surface.

Everything depends on which "language" mode of space you chose to depict the triangle in.

Assuming Volumetric depiction is motive then

triangle may be connection of any 3 lines formed by planes intersecting the sphere.

one cannot be the other you must be either either Euclidean or Geodesic or translate so ~~the~~ to the other view point can be interpreted.

FIGURE 13

Math 150—Classroom Assignment

Last time I talked about a triangle on a sphere.

How would you describe my definition? Does that definition make sense to you? Why, or why not?
Would you suggest an alternate definition, or do you believe none can be constructed? And why would it matter at all whether such a thing as a triangle on a sphere could exist, anyway?

Assuming Euclidean depiction is objective — *How do you mean this word?*

that which supperimposed onto a flat surface from the spherical surface, would form a triangle on the flat surface. — *This is a pretty good intuitive definition of triangle.*

Everything depends on which "language" mode of space you chose to depict the triangle in. — *How are "language" and "mode of space" related?*

But aren't WE doing 2-D stuff?

Assuming Volumetric depiction is motive then

triangle may be connection of any 3 lines formed by planes intersecting the sphere.

"One" what?

one cannot be the other you must be either either Euclidean or Geodesic or translate so ~~the to~~ the other viewpoint can be interpreted.

A lot of what you say here makes intuitive sense, but I'm not always sure of your meanings. We'll try to get more precision as the course develops.

FIGURE 14

her confusion. By phrasing this comment as a question, we are asking her to think more fully about one of the central points in her assignment, and are emphatically *not* telling her "what the answer is." Our goal is to guide her to the discovery of how to define the word "triangle" on the surface of a sphere, and she is aiding us in understanding that we will need to design future discussion and assignments to aid in achieving this goal.

We now see that, for instance, until she understands such differences as those between "in" and "on," and "two versus three dimensions," she won't be able to comprehend the concepts of spacetime and curvature that are going to be coming later in the course.

With this assignment, this student has shown us that she is confused, but she's confused in the same way a sixteenth-century mathematician might be confused about these and similar concepts. "What's 'straight?'," the mathematician of old might ask, "and how does it differ from 'curved'? And, how can I try to measure the difference?"

There are patterns of critical thinking going on in this simple draft. This student is thinking aloud; she is imagining the problem of "translating" the Euclidean triangle to the sphere by first visualizing the translation, and then explaining that visualization. When the mental image doesn't quite take logical shape for her, she admits "everything depends on which 'language' mode of space you chose," that is, the axiomatic system "you" (i.e.,"me," the original definer of the word triangle) have described in words (i.e., my "'language' mode of thought"). By writing this passage, she has shown us that she has not yet been able to narrate logically what a triangle on a sphere is. And with her writing she has given us some good ideas as to what questions we should ask to get her to a more complete understanding of the topic, and what assignments we could formulate to help her achieve this understanding. This is a lot to get from what has been described by one professional as "one dumb assignment—I can't believe you can call this writing!"

10

Learning from the Cognitive Faculty

On your way back from lunch, you pass the dean, who spends her "other life" as a faculty member in cognitive studies. "I hear you're working on a writing text," she says. "I hope you use some of the ideas from cognitive theory; I've found they really improve my students' papers." Then she adds, "Let's talk about it some time. I'd much rather do that than read more tenure files."

Papers, we mentioned in Chapter 7, may be categorized as *personal, informational,* or *argumentative.* This type of cataloguing helps the writers define their own strategy. The writing faculty have used this particular categorization to help students identify their own approach and thus write higher quality manuscripts.

There is another way to look at papers. This method emphasizes the subject matter of the essay; it asks about the level of sophistication of the manuscript. At the same time, it provides a template for types of questions we might ask. Such a classification, called a *taxonomy* of cognitive levels, has been suggested by various authors, for instance by Benjamin Bloom [Bl] and John Chaffee [Ch]. We will use here Bloom's taxonomy, mainly because it is so well known, often used, and fits the mathematical model better. According to Bloom, a paper can test for *knowledge, comprehension, application, analysis, synthesis,* or *evaluation.*

The most basic question tests for *knowledge*: "Define the word limit." Such a question is good but not deep; it gives some idea of whether the student has tried to absorb some essential information from the course. It is not dissimilar to examination questions asked on our standard tests.

A somewhat more sophisticated question along the same lines might be the one we used earlier: "What do *you* think the word limit ought to mean?" Although just a personal essay, this assignment is more likely to test the students' *comprehension* than is the previous question. By using their own words, the students will show whether they have absorbed the idea of limit; and whether the concept is intelligible to the students.

If the concept *is* intelligible, it is now time to see if the student can *apply* the concept. In a first-semester calculus course, we often ask the following argument question, "If you were trying to define 'limit' in three dimensions, given what you know up to now, how might you go about it? For instance, how would you [the student] cope with the many different ways you have to let a function approach a given point?" At this stage, the student is being asked to use, in a meaningful way, information he or she has been taught.

Notice that, at the application stage, the format of the question becomes important; we may have to help the student along by suggesting possible lines of reasoning.

The next type of question is the one that asks for some *analysis* of the concept. In the second half of the above question we came very close to asking for an analysis of limits in three-space, but we can make it even more specific: "Use words and pictures to compare what happens to the function $z = xy/(x^2 + y^2)$ as x and y head toward zero along the two paths $x = y$ and $x = 0$. Then try some other paths, and formulate some conjectures as to what might be going on here."

Such a question asks the student to "really get to the heart of" a major course topic; to cut it apart and carefully investigate its components, to subject the topic to our analysis. This process is not dissimilar to what many of us have done in real analysis courses: "What is it that makes the intermediate value theorem hold? Ah, yes, we need continuity of the real-valued function and connectedness of the domain."

An even deeper question is one which requires *synthesis*. Synthesis papers fall into two main types: low level and high level. Here's a low-level query which requires more than freshman calculus for a response: "Compare and contrast the various definitions of 'limit' which have been given by mathematicians over the centuries." At a higher level, the student might be told to "Discuss the factor which led from Cauchy's definition of limit to the Weierstrass $\varepsilon - \delta$ formulation." To answer this last question, the student is going to have to know a great deal about the history of mathematics and about the theorems and examples that have shaped the changing definition of limit over time.

Synthesis, meanwhile, is a process whereby we put together disparate or seemingly disparate ideas or lines of reasoning into a coherent whole. "How did the ancient Greek philosophers' attitude toward the word 'infinity' affect their study of velocity; and how did Isaac Newton's attitude affect his?" "When Wallace Stevens, the poet, talks about 'The Idea of Order,' how does his definition of order compare with the mathematical definition?" Such questions can lead to argument papers with a high level of cognitive and mathematical sophistication.

Analysis and synthesis questions make for good major projects in writing-intensive courses; at the same time, they can "blow away" students who are asked them too early in a course.

The final type of question requires the student to *evaluate*. "What were the impulses which led people to be dissatisfied with the previous centuries' definitions of limit, and how did these impulses shape the next definition?"

Evaluation questions are most complex; a good evaluation paper can lead to a thesis, not to mention a lifetime of research. "What are the fundamental differences between Wallace Stevens's poetic idea of order, and the mathematicians'? What is gained and lost in each definition?"

By knowing the several levels of cognitive skills, we can design questions which can test students' understanding of the material presented in the course. We can ask them to respond at a level they can cope with, while formulating questions and projects so that they can grow into the complexities of the subject.

Now let's turn the tables: You are teaching algebra. A student says he couldn't factor the polynomial

$$10x^2 - x - 3.$$

He is showing that he does not yet have the knowledge of how to factor some quadratic equations. After you show him the solution, if you then ask, "How about if you try

$$15x^2 - x - 1,"$$

you are asking him for comprehension of the technique you just taught.

You might next ask him if the same methods can be used to factor a particular cubic equation. This requires the student to work at the level of application, and perhaps at the level of analysis.

When the student asks you if there's a formula to find roots of every possible equation, he is asking you for analysis and synthesis of the information *you* know about algebra. Then, when he says, "What's algebra good for, anyway?," he's asking for evaluation.

You, the good instructor, will be prepared to answer all the questions, at all the levels above, and to do so in a way that will make some sense to the student.

Exercises/Discussion Questions

1. Another student asks why you would want to find the roots of a quadratic equation.
(a) At what level is her question?
(b) How do you give her a short answer she will understand and accept?

2. A third student asks when you need to use a quadratic equation versus a linear one.
(a) At what level are you being quizzed?
(b) How do you respond?

3. Compile a list of questions at the six levels given in the chapter on one or more of the following topics: continuity; trigonometric functions; angles; vectors; algebra; geometry.

4. Make a list of questions your students ask in the next class; catalogue them by level. How would you raise the level of some of their questions without "losing the class"?

5. We have some reservations about some of Bloom's taxonomy, especially at the higher levels of cognition. Can you design some assignments which will show us a substantive difference between the cognitive levels of synthesis, analysis, and evaluation?

6. The Chaffee taxonomy emphasizes a different catalogue of tools. He suggests using language as a means of, among other topics:

Forming and applying concepts;
Relating and organizing processes;
Comparing relationships;
Reporting, inferring, and judging.

(a) Suggest some mathematics assignments based on Chaffee's categories.

(b) Chaffee's model is sometimes referred to as a "horizontal" model of cognition, while Bloom's is a vertical model. Discuss what this means.

Coda E — An Extended Calculus Assignment

Very few students who enroll in calculus will need to compute derivatives or integrals later in their lives. This is even disturbingly true of students from our client disciplines. Yet many majors and general college curricula require a certain number of mathematics courses. In order to justify the central role mathematics plays in undergraduate education, mathematicians often claim that the learning of mathematics goes beyond computation; it fosters critical thinking and analytic skills. In our introductory courses, writing assignments which require narrative and cognitive strategies help develop students' thinking and analytical skills.

Consider the following (very traditional) question from a quiz in first-semester calculus:

> Compute the x and y coordinates of all points on $x^2 + xy + y^2 = 3$ which have horizontal tangents.

In terms of Bloom's taxonomy, this question sits somewhere between knowledge, application, and synthesis. It tests the students' knowledge of how to do implicit differentiation. We gave this quiz early in the discussion of implicit derivatives, so the students had to apply what they had learned (the technique) to a problem they had not previously encountered (looking for horizontal tangents). Further, once the students realized that they needed to set the derivative equal to zero, they would have to back-solve to find the actual points; they had to synthesize their new technique with previous algebraic methods they had learned in high school. What might seem like quite an innocent question is really quite demanding.

Not surprisingly, most students successfully found that $\frac{dy}{dx} = -\frac{2x+y}{x+2y}$ using implicit differentiation. They could perform the computation. However, less than half could figure out how to use the fact that if $y = -2x$, then $\frac{dy}{dx} = 0$, to look for the points in question. Only two successfully substituted and solved to find possible points: $(1, -2), (1, 1), (-1, 2)$, and $(-1, -1)$. Neither of these two checked to find out that only the first and third solutions actually worked. The question was admittedly challenging, but nonetheless, the results gave us impetus for further assignments.

The course in question was an experimental section using graphing calculators. In order to improve the students' ability to apply and synthesize their knowledge, and to improve their ability to effectively use the calculator, it seemed like a good idea to combine implicit differentiation with the graphing capabilities they had available. On Monday, after going over how to solve the quiz problem, and giving a short lecture on orthogonal families of curves, the class was asked to work on the following questions:

> 1) Figure out how to use your graphing calculator to sketch the curve implicitly determined by $y^2 + x^2 - x = 6$. Use this graph to estimate points where the tangent is horizontal.
>
> 2) Repeat question 1 for the curve given by $y^2 + x^2 - x = 4$. What goes wrong in your calculator's graph?
>
> **Due Wednesday:** Write a short note to a friend of yours in another section of calculus, explaining how to graph implicitly defined functions on a graphing calculator. Your note should be clear, it should indicate a specific difficulty that might arise, and it should contain an example *not* from parts 1 or 2.

(In case you don't have a graphing calculator handy, when you combine the graphs of $y = \sqrt{6 + x - x^2}$ and $y = -\sqrt{6 + x - x^2}$, you get a nice circle with x-intercepts of -2 and 3. However, in the second question the x-intercepts are irrational, so there are "gaps" in the picture sketched by most graphing calculators.)

These essays provided excellent material for Friday's class. Three of the difficulties mentioned in student essays were written on the board, and the class was asked to comment on them.

> "This method works fine if you've got a y^2 and no other y terms, but otherwise you can't really use your calculator."
>
> "The graph has gaps that aren't really there. You shouldn't use your calculator because there's no way to know if you've got the right answer or not."
>
> "The only difficulty is for calculus. Now that I know how to get these curves on my TI-85, there's no need to compute derivatives. I can find maxs and mins faster by looking at the graph."

These questions indicate that the students were starting to analyze what could and could not be done with their new tools. They were beginning to ask: What good is implicit differentiation? Why should we use a graphing calculator? And good instructors should then ask themselves: If the students can't answer these questions, what's the point of moving forward with the class?

The first statement really got the ball rolling. A few students had realized they could use the quadratic formula to graph a number of interesting curves. For example, if they wanted to graph the curve from the quiz ($y^2 + xy + x^2 = 3$), they would just need to graph the curves

$$y = \frac{-x \pm \sqrt{12 - 3x^2}}{2}.$$

A few people pointed out that equations "quadratic in y^2" (like $xy^4 + x^2y^2 - x = 1$) could be handled in a similar manner. Ultimately we did decide that while this was a useful and interesting bag of tricks, there were a lot of polynomials in two variables that their calculators couldn't handle. On the other hand, there was no such polynomial they couldn't implicitly differentiate! We also discussed how one could use facts about the derivative to help pick viewing windows on their graphing calculator, and to predict where errors (= points of vertical tangency) might occur.

After getting back their essays, the students were asked to rewrite them. This time they were to imagine that they were writing a new chapter for their text titled "Implicit differentiation and graphing calculators." Their job was to produce an example to be used in this chapter based on their previous example, and in the in-class discussion. The first assignment and the discussion were part of the process that prepared the students to create their examples. The question, "What goes wrong in your calculator's graph?" (from the first assignment) was a prewriting assignment for their first draft; the classroom discussion was prewriting for their final product. The examples were excellent, polished, and varied. Some students concentrated on techniques for graphing, others on errors and vertical tangents, and some on using the calculus to inform the picture produced by their calculator.

Of course, all good experiences should be reinforced. So on their final examination they were presented with a question connecting our extended discussion of implicit differentiation with a later discussion on integrating the capabilities of calculators with calculus techniques to produce accurate graphs of complicated curves.

> a) Find $\frac{dy}{dx}$ for the curve implicitly defined by $y^3 - 3x^3y + 9xy = 1$.

b) Estimate the points of horizontal tangency correct to three decimals. You will need to use your calculator in this computation. Full marks will be given only if you fully explain your work.

This is quite a challenging problem. The students had to correctly compute $\frac{dy}{dx}$, then do some mildly complicated algebra in order to see that horizontal tangents may occur when $y^3+6y-1=0$ or $y^3-6y-1=0$, and then use their calculator to estimate these points. Most importantly, they also had to be able to justify the steps they were taking, in order to get full marks. While not everyone got this question, they did much better on it than they had on the earlier (and much more elementary) quiz problem. They were beginning to move away from trying to learn mathematics only by memorizing particular techniques.

Major Projects

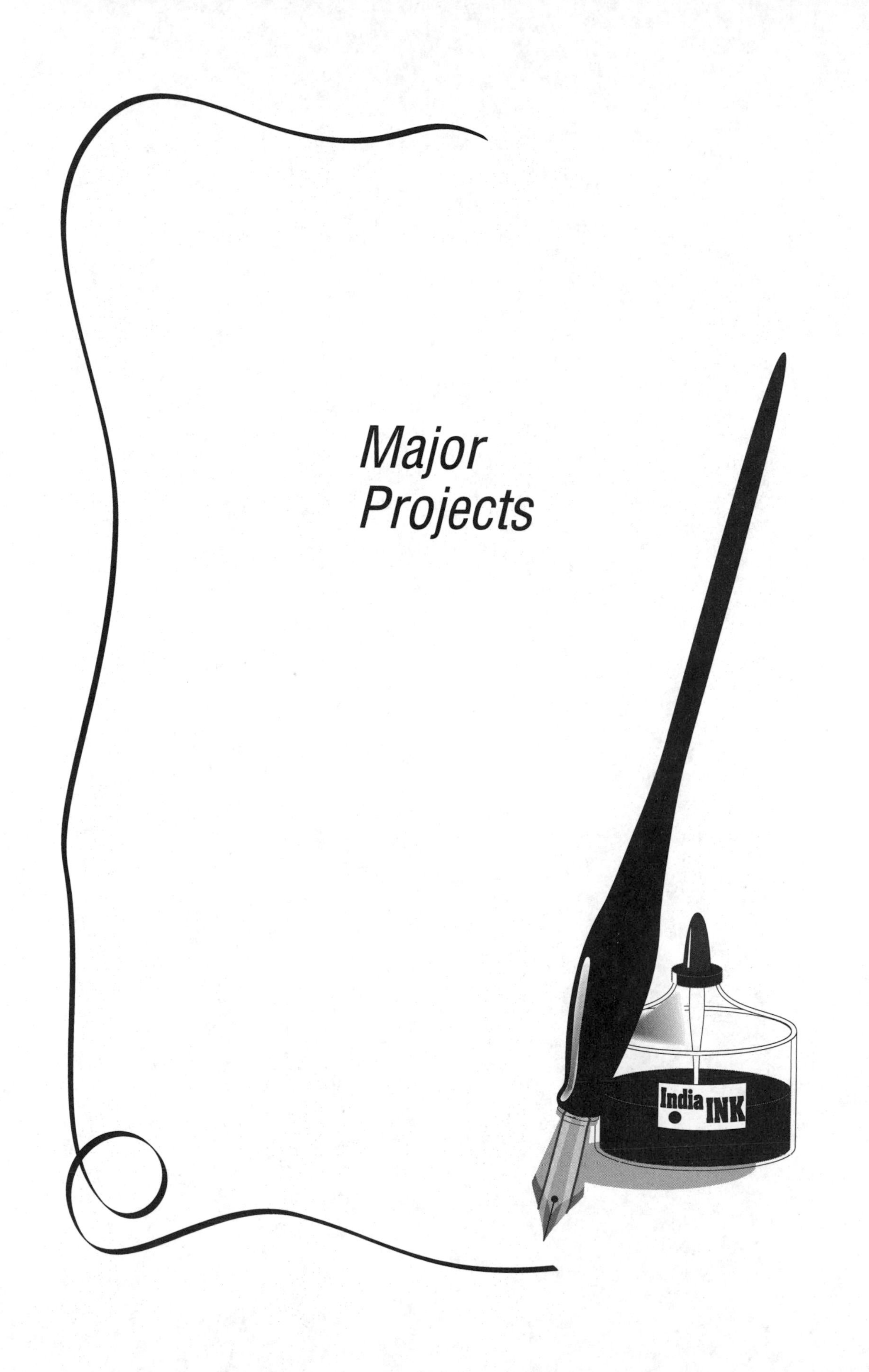

11

Talking Through the Problem — Conferences

You pass by the room administrator's desk. "We need to know when to schedule your final exam." "Final exam?", you ask. "It's only October." Then you realize—you assigned "major projects" instead. And, you have to start thinking about them *right now* . . .

Somewhere around week eight of a 14–week semester, we ask students to write us a short note. The first question we want them to answer is, "What are you thinking of doing for your final project?" The second question is, "What materials and help do you want from me?" Then, after we have had time to absorb the students' answers, we call them into the office for 20–minute individual conferences. We ask each student to pare down his or her topic, if necessary. For instance, if a student says, "I'm interested in writing about the effects of geometry on architecture," we will ask "What types of architecture? Can you give me more specifics about how you think 'geometry' enters into the aesthetic considerations of the buildings or landscapes you want to discuss?" Then we will suggest some readings, sometimes from our own libraries, as references for the student to consult and use for the project paper.

A few days to a week later, we meet again with the student to get a verbal description of the thesis and to find out if the references were relevant. In this way we find out whether the student:

1. Actually has a reasonably pared down thesis;
2. Is working steadily on the project;
3. Has begun reading the reference list.

At the same time, we can begin to acquire some understanding of the student's work. We can read the references, and make suggestions as to, for instance, experts who might be able to help the student.

Here's an edited version of an example, culled from a recent year's work:

Prof: Have you gotten any ideas as to what you'd like to do your project on?

Student: Well, we've been talking a lot about Christopher Columbus and how he navigated, and I thought I'd try to find out some more about how his trip was done.

Prof: That sounds like a good topic. You need to be more specific, though. Where do you propose to start?

Student: Well, I've been reading more of the [Samuel Eliot] Morison book [Ms], and there's a whole chapter in there about how Columbus navigated; in fact, I think that's the title of the chapter. I guess I'd use that, plus maybe Columbus' diary.

Prof: Sounds good. Now remember, the topic should have something to do with geometry, so you have to talk about that kind of thing. . . . (Here we're not too specific as to what we've suggesting to the student; we want him to find the topic himself.) Do some reading, and come back when you're ready; say, in a week, and let me know what you've found out—'cuz, I'll want to hear how you're coming along so you don't get too far out on a limb without help. . . .

Student: O.K., thanks. Then that topic's all right?

Prof: Yep—see you in about a week. We'll figure out when in class.

The next week, after class:

Student: Can I talk to you some time about my project?

Prof: Sure. When do you want—or is right now all right? Generally speaking, what's it about?

Student: Well. . . now's O.K. See, I noticed that Morison says that Columbus used "taxicab geometry" in part of his navigation, and I thought I'd write some about that. . . I mean, I've got a lot of other stuff about how Columbus navigated, but this one's really geometry . . . and so, I thought I could explain how taxicab geometry works. I mean, I found this book, and. . . .

Not all conferences go nearly as easily. Here's a rough, but reasonable account of another student's experience:

Prof: So, have you thought about what your topic will be?

Student: Well, yeah, but I didn't get any good ideas. . . . What do you want me to do? (Here, the student is hoping we will tell him what to do. We don't accept the bait.)

Prof: Well, of all the things we talked about in the course, which ones appealed to you most? Did you like the design stuff. . . .

Student: Nah. . .

Prof: . . . or maybe the surface theory . . .

Student: But I don't see what I can do with it, and besides, it's too hard.

Prof: How 'bout higher dimensional ideas.

Student: Nope (Emphatically).

Prof: . . . or art? Poetry? The material on hyperbolic geometry?

Student: Uhhh. . . (Pause).

Prof: Well, let me look at my bookcase. Here's something on the Platonic solids; we didn't really discuss them in class. You could work up something on that topic. Or, how about the golden ratio; I have a couple of books on that?

Student: What's the golden ratio? (First inkling of life. . .)

Prof: It's about (etc.). Here. Let me lend you this book—and take the one on the Platonic solids, while you're at it. You can look over both of them, and maybe we can talk next week after class about whichever topic seems to catch your eye.

Next week:

Prof: Have you looked at the books I showed you?

Student: Yeah. The golden ratio one seemed kinda' interesting. I thought I might do something on that.

Prof: O.K., that sounds good. But that's a big topic, and I think you're going to have to trim it down. How about if you come to my office and we'll see what part of it interests you most. . .

Neither of the above dialogues represents the norm, of course, but these relative extremes should give some idea of how the conferences proceed.

If conferences are too time-consuming, we ask students to write half-page descriptions of their projects. We can then give short written responses of the "This sounds good—go for it," or "Too tough; sharpen your thesis" variety. We can also use class time to have students share their proposals and approaches.

A few generalizations can be made about conferences: The instructor should think beforehand about the following helpful hints one can give:

- What is a reasonable range of topics that I know well enough to actually be able to help the students positively?
- Which classroom ideas will be the best for the students to try to extend into meaningful assignments?
- What books, software, etc., do I have or can I suggest to be of substantial use by the students?

Meanwhile, the instructor should be considering the students' perspective:

- Do the students find the atmosphere of the conference unintimidating?
- Do the students really value a particular project? This last is particularly important, for, if the students feel that the project is one they have discovered, then they are more likely to work hard on it, even if the instructor doesn't consider it particularly interesting, or doesn't know that much about it.

Later conferences are arranged, as needed, to discuss students' papers, as well as problems that have come up, and to keep projects on schedule. We offer to "look at first and second drafts" to make general suggestions; most students take us up on this. From their viewpoint, we are telling them what we want, and helping them get a better grade; from ours, they are checking their sources early, getting a decent draft done in time, sharpening their theses, and sticking to the task. The fact that the students get a clear message as to what is expected of them in their major papers is important in settling their minds on the thesis topic, and it also diverts them from spending their thinking time trying to guess at "what my instructor wants."

Exercises/Discussion Questions

1. Think of five topics in the course you are currently teaching that could be material for a project. What books or other materials would be useful to your students in successfully completing such a project?

2. Suppose you are teaching calculus, and a student gets interested in the fundamental theorem. Think up two possible projects, one having to do with a historical perspective of the topic, and another dealing with some application of the fundamental theorem. What are your references? What kinds of "applications" will the student be able to understand sufficiently to actually get something out of? What kinds of applications will the student really understand at the end of the project, rather than having simply copied them out of a book?

3. In a differential equations course, you have assigned a project of finding "real world" applications. A student tells you he has heard that the equation he is studying has "a lot to do with electrical theory, and it sounds interesting," but he "doesn't know anything about electricity." You don't either. He still wants to try the assignment. Now what do you do?

12

What "Guzinta" a Good Major Project

You've successfully negotiated half the semester, and you're finally beginning to feel secure about teaching an expository course—when suddenly a nervous student, one of your best, appears at your door. "Professor," she says, "I don't have any idea of what you want for this final project you gave us." "Neither do I," you blurt out.

The best major project is a group effort—between you and the student at least, perhaps also a librarian, another "favorite" instructor, a fellow student, and who knows who else.

What does your student like? What is the student's specialty? At what level of maturity is the student? What has the student been writing for you up to now? How successful has that writing been in getting the student to understand mathematics?

It is no secret that you will get much more, and much better, writing on mathematical topics if students have a topic that they see as being "really relevant." So, when the students comes to you, ask them about their major, about the other mathematics courses they have taken, about which topic they have found most appealing so far, about what they want to do when they graduate. But we discussed this in the chapter on conferences, so let's go on to the question of what's a good final project.

Some students like the "science fiction" aspect of certain mathematical topics. One of them might write a paper based on Edward Abbott's *Flatland* [Ab]. The student might discuss some of the differences between two- and three-dimensions, and you can encourage the student to find two or three differences that Abbott does not discuss. Or, you can ask the student to write a similar story about, say, a sphereland (this has been done, by Rudy Rucker [Ru] and Dionys Burger [Bu]), or torusland. Or how about a Klein-bottleland?

A more advanced project along similar lines would have a student try to envision a toroidal Earth. How would gravity operate? Which way would water flow? How would maps be drawn? How is air travel done? And, why aren't there any toroidal planets (that we can see) in the universe?

An even more advanced project might discuss the geometry of the universe, locally and globally. The student could talk about curvature and mass, about black holes, about gravitational lenses, and how they affect the geometry of the universe. Some other references for the topics are Petit [Pe] and Misner [MTW].

For a student with a little background in calculus, a discussion of the changing limit concept over the centuries is a possibility. Another is the question of who were precursors of Leibniz and Newton, while a third is a study of the infinitesimal.

A somewhat more advanced student might write a paper discussing the difference between calculus in two dimensions and three. Of course, as the level of such a student increases, the number of topics that can be included in that student's paper can increase—from, for instance, a simple description of the distinction between ordinary and partial derivatives, through the discussion of continuity and differentiability in three-space, to an understanding of the Cauchy–Riemann condition and their impact on infinite series. Edwards's book [Ed] is useful here, as are Stillwell [Sw 1, Sw 2] and, for historical purposes, Struik [Sk].

If you have a number of art or architecture students, you can suggest that they work on projects which combine design, aesthetics and geometry. What is the best design of an auditorium for acoustic purposes? How can mathematics be used to analyze traffic patterns in buildings, or on roads? What kind of geometry does the camera lens have? Is there really a most pleasant shape for a room? An analysis of the golden ratio is a popular topic.

Students of literature or philosophy can also have projects which appeal to them. Some of ours have analyzed poems by Wallace Stevens [Sv], A.R. Ammons [Am], and Howard Nemerov [N] by looking at the geometric words in them, drawing pictures of what is being represented in the works, and then using those words and pictures to interpret the poem. Another student, who was interested in Plato's banishing of the artists from his ideal city in *The Republic*, made a scholarly comparison between Plato's attitudes toward art and mathematics.

The history of mathematics can be a source for many projects in a writing course. I have had students discuss the acceptance of the concept of complex numbers, compare current methods of proof of the quadratic formaula with earlier proofs given in antiquity and the middle ages, write on Galois' life and outline Galois theory, and discuss various aspects of the history of hyperbolic geometry. Useful sources are Bell [Be], Boyer [Bo], and Struik again.

Statistics is another field which can be mined for projects. Two ideas are: a statistical analysis of feeding patterns of codfish when subject to various amounts of pollution; and, an analysis of the success or failure of the Canadian government policy of moving Newfoundlanders from outports to cities after the province joined the nation.

Statistical data from the most recent United States census is now on line, and can be a gold mine for student projects.

Once you get the idea, you can design numerous projects of your own which will fit your personality and those of your students better than those listed above for my students. Start with your field—is it algebra? What is a basic topic in that field? Is it linear transforms, say? Or is your student an engineer? Maybe that student would like to find out how linear transforms are used in computer graphics.

In his geometry course for people who "know no geometry," Tom had a student who did all the assignments well, but clearly showed much more interest in those that had to do with fine arts. In one of her early papers, she described how art and geometry interacted, and a later paper ("describe a work of art at the local museum, starting from 'geometric words' we have been using in our earlier discussion of design") dwelt at some length on a particular cubist painting. From classroom and conference discussion with the student, Tom knew that she intended to be an art major, and that she had a fair amount of knowledge of the Cubist movement. Meanwhile, the class in general had, by trying to understand the Greek mindset, become interested in the axiomatic method. Tom suggested, and the student agreed, that she would profit, both mathematically and intellectually, by attempting the following final project:

> Compare the principles of cubist painting with the axiomatics of mathematics. What are the underlying assumptions of cubism? In what way is a painting like a theorem?

After the student had completed the project, she and Tom both agreed that it had been interesting and worthwhile. More importantly, it was a topic that she could continue to think about and grow into over the years.

Another project evolved for a student who saw herself as an artist and writer, and felt sure that no aspect of mathematics was relevant to her life. She often said that she only took the course because she heard there "really wasn't any mathematics in it." She did, however, participate in classroom discussions of the early Greek philosophical motivation toward abstraction, and became interested in what this motivation had to do with their way of doing geometry. Eventually the student wrote her final project on the difference between Plato's and Aristotle's beliefs concerning art and mathematics, and why Plato banned artists from his ideal city. And, it should be mentioned that this student took calculus the next semester.

At a more advanced level, students can also be induced to find interesting projects even more quickly. Some instances:

In history of mathematics, a young man looking for a project mentioned that he was also taking complex analysis. He said he was getting lots of theory, but not "the big picture." So he designed a paper on the evolution of notational schemes for complex numbers, and how that evolution contributed to the eventual acceptance of complex numbers and complex analysis.

Another student, majoring in mathematical education, wanted a project which he could use in his high school the next year. He eventually wrote on the history of trigonometry.

Final projects should meet the level of ability of the student. One of the people in an "elementary" class was able to handle the idea of explaining "baby Morse theory," which relates the number of pits, peaks, and passes on a two dimensional surface to the topological concept of the Euler number. Another student drew graphs of conic sections in so–called "taxicab geometry." A third gave an exposition of current theories about Olber's paradox, which is the question of why the sky isn't light all the time. A fourth wrote about how she would explain the basic ideas of two-dimensional hyperbolic geometry to high-school students. Yet another did an analysis of the numbers of vertices, edges, faces, hyperfaces, etc., for higher and higher dimensional cubes.

In history courses, people have discussed Diophantus and his effect on algebra; the artistic origins of projective geometry; which parts of calculus were "discovered" first by Newton, and which by Leibniz; the ever-popular origins of non-Euclidean geometry; and the beginnings of Indian mathematics.

Exercises/Discussion Questions

1. Do you agree that a paper on Cubism is really a valid assignment in a mathematics class? If you do agree, what similar topics might you suggest? If you do not agree, how would you revise the assignment until it became one that you did find valid?

2. In this chapter we have left out some details about items that students want to know but instructors will have different feelings about. For instance, how long should a "final project" be? And how many references should a student have?

3. Conjure up two final projects which you would like to see a student in an elementary class carry out, complete with a short, quick list of possible references that student might use (Yes, you can do this—it's not as hard as your thesis!). Now do the above for a more advanced-level student.

4. What subfield of mathematics interests you? Think up a few possible projects for students at basic, intermediate, and advanced levels. What kinds of questions can you expect from these students as they proceed? How much help should you give them? What texts or references will they (and you) find useful?

5. Do you find all of the projects in this chapter "acceptable" for your class? Why or why not? Given that you are teaching a specific class at a particular level of ability, what criteria do you have for your students and their projects? (Note that the better you can articulate these criteria to yourself, the more likely you are to be able to do so to your students when they ask about the course and assignments.)

13

Presenting Major Projects

You and your students have been working hard on their major projects. In individual student conferences you have seen some very exciting work being developed. How can you let everyone in the class know about the good work that's being done?

We consider talks given by the students to their classmates an important part of the process of creating papers. In classes such as real analysis, abstract algebra, or geometry, where major projects fit easily into the syllabus, we often require our students to make classroom presentations based on the work in their papers. Each presentation is 10 to 15 minutes long, given in a conference format. If the "conference" occurs close enough to payday, we provide doughnuts, coffee, and juice. (We prefer to teach in the mornings; the menu might need to be altered for an afternoon or evening class.) At times we have the students write abstracts which are photocopied into a schedule for the class's "conference"; sometimes, when the work has been excellent, we publish a "conference proceedings" containing the best papers.

Although it takes up some class time, the experience of presenting is good for your students and produces stronger papers. Students get to articulate their ideas and to verbalize what they have learned. They benefit both by being a presenter and by being in the audience. Further, by going through the process of creating a short, clear and interesting talk for their classmates, they better understand the mathematics in their papers. Since you often want them to write their papers as if their classmates were their audience, making the class the *actual* audience for the talk will get them to write their paper at an appropriate level. Even more than writing, talking seems to force people to impose a narrative structure on their work.

Narrative elements are found not only in fiction. Almost any paper needs to present its information in a coherent and interesting manner; the lack of narrative leads to dull, lifeless writing. When we don't "see" the audience, it is easy to forget to let the reader know why the paper is interesting and important.

Students also benefit from listening to peers present their work. They can ask questions of each other, challenge (in a nice way) various ideas, and offer further exporation on the topics. They gain insight about how to present mathematics; and the small amount of competition introduced by hearing about other people's work could be sufficient impetus for writing "just one more draft." Finally, they will benefit by listening to a dizzying array of mathematics presented in a non-threatening environment.

Let's look at a couple of plausible situations which are amalgamations of actual class experiences in a sophomore-level geometry course.

Toward the end of the semester each student was asked to do a small amount of research into some aspect of geometry not covered in class. *The Penguin Dictionary of Curious and Interesting Geometry* by David Wells [Wd] offered a good primary source for topics. The students began by reading an entry in the dictionary and then looking for more information on that topic. For their paper they were to locate or create a proof of the result, discuss the history of the topic, or connect the topic to concepts covered in this or some other mathematics class. In class, as part of the prewriting process, each student declared what topic he or she wanted to work on (to avoid duplication) and then in individual conferences the student and instructor discussed ideas, outlines, etc. In a follow-up conference the student and instructor discussed a first draft of their paper and an outline of the student's talk. (If we didn't have time for this conference, the class could "workshop" this in groups.) In this second conference the student was essentially giving a practice talk. The student was presenting material to us, and we indicated what aspects of this presentation were clear, interesting and effective and which were not. One week after the talks, the final draft was due.

Now for two actual examples from two different students:

Erma was an outstanding student who was clearly going to receive an "A" in the course, barring some unforeseen act of God. In typical fashion, Erma had read, marked, and digested her article days before the first conference, and, when we met, she had already consulted further sources. In conference, she declared that she had planned to write about the Möbius band, but that she hadn't found the material in the dictionary that interesting. Instead, she decided to write a description of how to make a Klein bottle out of two Möbius bands by joining them along their boundaries. After a few minutes of talking, it was clear that she had a good idea of what the Möbius band is, and she seemed to be very interested in the non-embeddability of the Klein bottle and the projective plane in three-space. We asked her to work on the following problem as an exercise:

> The standard picture of a Möbius band has the boundary of the band essentially running about on a vertical cylinder. Is there a way to embed the Möbius band in $\mathbb{R}^3$ where the boundary circle is contained in a plane $\mathcal{P}$ and the rest of the band is on one side of $\mathcal{P}$?

Of course, it is impossible to do this in $\mathbb{R}^3$; if it were possible, we could place another copy of the Möbius band on the opposite side of $\mathcal{P}$, join these along their boundaries (in $\mathcal{P}$), and we would have an embedded Klein bottle! This is a hard problem, but Erma was a strong student, and one who had learned quite a bit of material. Perhaps, we thought, by "solving" this exercise she would develop an interesting approach for her paper.

Erma stopped by the office a couple days later, having discovered the major difficulty in the problem. She now had a deeper understanding of what the construction of the Klein bottle from Möbius bands says about the connection between the Möbius band and the Klein bottle. After checking with me, she decided she was comfortable with this exercise *not* appearing in her paper, which had moved far from her original inspiration. She decided instead to write about playing tic-tac-toe on a Klein bottle.

The story isn't flawless, however. Erma's penultimate draft described various winning positions for $n \times n$ tic-tac-toe on the Klein bottle. The work was quite involved, but it was a dry paper; her talk consisted of essentially reading out her section on certain 5×5 cases, and her audience was bored and somewhat confused. Although the previous work Erma had done was excellent, and in her talk Erma had concentrated on a well defined topic, her presentation of material in the talk and in the paper really wasn't very good.

This example presented a few dilemmas for me as an instructor. How could I constructively inform Erma that she could do a better job? I thought I might mention narrative structure and remind her that she was telling a story, but that she hadn't told it in a way which would capture her classmates' attention. I decided to ask her which of the other student talks she had enjoyed most, and then to start discussing what was different between her presentation of material and the presentation she had most enjoyed. Did Erma have too many details? Not enough motivation, i.e., a good introduction?

As part of my discussion with Erma I would emphasize tone. Although she was writing a mathematical paper, she did not have to present the material in a dull manner. I'd praise her for her clarity and effort, but indicate places where she could write less formally, and better maintain her reader's interest.

It is important for me to note here that, when Erma engaged her audience, she was teaching, and as she taught, she was also learning. She was reinforcing the ideas she had already formed, and discovering new ones in the process.

Ernie was a different case all together. He had been struggling in my course; in the first conference it became clear he hadn't yet settled on a topic. He even explicitly asked if he could work on the "easiest topic" available. After a short discussion I found out that Ernie enjoyed the combinatorial arguments from the start of the course, but he wasn't very confident about the more recent material. I suggested that he should look at Pick's theorem because it is very combinatorial, and it has a number of practical applications.*

The situation was somewhat delicate during the second conference. Ernie had read about Pick's theorem, and he had written a rough draft of his paper which was disquietingly similar to the the entry in *The Penguin Dictionary*; I was worried about plagarism and that Ernie hadn't yet connected with the topic. I noticed that Ernie had quoted the conclusion of the entry in *The Penguin Dictionary*: "Pick's theorem is equivalent for a plane map to Euler's relationship for a polyhedron: vertices + faces = edges + 2." He had not, however, discussed Euler's relation anywhere else in his paper. "What do you know about this Euler relation?" I asked. Happily, Ernie knew quite a bit. He had checked out a couple of relevant topology books from the library, but for some reason none of this material was in his paper. I discussed Euler's relation with Ernie, giving him a quick tutorial on points which he wasn't clear about, and after a while I was satisfied that he really did understand a fair amount about, and was interested in, this theorem. "Take a look at Lakatos's book *Proofs and Refutations*," I told him, "It talks a lot about the history of the Euler relation and it's long, but fun to read. See if you can't use that and these other books to add in more details on this Euler relation and try to make the connection

---

* Pick's theorem gives a formula for the area of a polygonal region whose bounding vertices are vertices in some lattice in the plane. If I is the number of lattice points inside the region, B the number of lattice points on the boundary of the region, then the area is given by the formula $I + \frac{1}{2}B - 1$. Pick's theorem is often used to make quick estimates of the area of irregular pieces of land. For instance, forest fires rarely burn in a regular shape, so the number of acres burned in a forest fire is often estimated by using topographic maps and Pick's theorem.

with Pick's theorem more clear." And since the deadline was drawing near, I added "Stop by tomorrow and let me know what sort of progress you are making."

Ernie did a good job on his overnight project. He liked Lakatos's book, and between that and his other texts he understood at least one way of proving the Euler relation. As he walked out of the office, he declared, "Maybe I'll even come up with a direct proof of Pick's theorem."

Let's get right to the bad news: Ernie's penultimate draft was a polished version of his rough draft; it was simply a revision of his source article with a bit more information on the Euler relation added in, but without making the connections clear. The good news came during the talk: Ernie didn't get too far into his talk before someone asked a question. Ernie gave a beautiful and insightful answer. Then another question came, and he gave another great answer, and so on until he was out of time. Even though his paper was not convincing evidence that Ernie understood his topic, his talk convinced me that he had gained some insight. In fact, without the talk there would have been no significant signs that Ernie has understood the connection between Pick's theorem and Euler's relation.

At the end of class I spoke with Ernie. "I'm really sorry about the talk." he said. "They just kept asking questions and I couldn't really even get to the stuff on Pick's theorem." I told him to pause a bit and remember the other students' questions. Most of them occurred because Ernie jumped right into the middle of his topic without giving clear definitions at the start and because he began using the Euler relation without explaining its connection with Pick's theorem. I pointed out that when we tell a story we start by describing the location and the characters. In other words, Ernie didn't do a good job laying the groundwork. "But," I added, "you gave some great answers. You've clearly got some real insight into what is occurring in Pick's theorem and in the Euler relation. Think about how you answered your friends' questions in class today, and then see if you can't use that in your final draft to explain this material in your own words so that someone else would have those questions answered in the paper." Essentially what I told Ernie was to write in his own voice; I also told him that he could do a very nice job of explaining this material in terms he is familiar with. I was asking him to think through the process of elucidating mathematical ideas to other people; what is unclear, and which examples and intuitions can resolve that difficulty?

Both of the stories above highlight the connection between student presentations and their papers, how giving talks can improve papers, and how talks can indicate weaknesses or strengths in lines of reasoning you might not otherwise notice. They also indicate that instructors need to work hard if they want to ensure that the student presentations will go well. Thankfully, preparing students to present their work is not too dissimilar to helping them construct their papers. Many of the issues revolve around students having difficulty understanding the proofs (can they follow them in special cases?), or wanting to do too much (you can't recount all of calculus or algebraic topology in a ten-page paper!). Whether they are preparing papers or presentations, we want our students to communicate their ideas effectively; it's just that the medium of communication is different.

Often the need for narrative elements is more clearly pronounced in talks than in mathematics papers. Almost everyone reading this chapter will have been, at some point, sitting in a lecture, wondering why the speaker is dragging the audience through certain details. When writing or speaking, one needs to clearly introduce the issues involved and stimulate interest in the material being presented.

As a rough guideline, we usually recommend the following tips to our student speakers in 10- or 15-minute talks.

- You know more than you can possibly say. Decide what is the most important concept you want to get across before preparing your talk.
- Probably you won't have time for elaborate proofs. Can you illustrate your main concept by a particular example which is easy to describe?
- Spend a lot of time preparing your introduction and conclusion. Make sure your introduction will be interesting to your classmates. Make sure your talk supports your conclusion.

We often find that we are very good at advising students about their talks. We all probably give 10 or more lectures every week. One thing we should watch out for, though, is trying to turn our student's talk into a presentation we would give. When we are meeting with an Erma or Ernie in our office, we should try to make sure we are clear as to what *they* want to say, and then help them design an effective short talk to get this message across.

In-class student conferences are very popular. The students pay attention to the speakers, and the speakers present a dizzying array of mathematics related to the particular course.

Because they have had to articulate their ideas, the class seems to achieve real learning during the talks, even though they are not tested on the material, or perhaps, even *because* they will not be tested on this material. It may well be that the listeners, witnessing their peers present their papers, get more out of the experience than the presenter. They get to see all of these marvelous bits of mathematics as well as several different approaches to presenting mathematics.

It's hard to guess the net effect that presenting papers had on students like Ernie and Erma. Certainly it improves their papers, and learning to organize and present difficult material is good practice for anyone. Having talked with former students, we've also noticed that they often remember what occurred during their talk even more clearly than the contents of their paper. When we have students present their work in class, not only are we offering them an audience but we are encouraging them to develop their voice; we are allowing them to be the authority on their particular topic. This is often a quite empowering event.

Exercises/Discussion Questions

1. Most teachers are very worried about coverage; they aren't completely confident that they will find the time to do all the things they feel they need to do before the course ends. Having students give presentations on their work could take up a full week of class time. How can you make this week a valuable learning experience for the student as audience, not just for the student as speaker?

2. You're teaching real analysis, and one of your students is working on everywhere continuous, nowhere differentiable functions for her final project. When she says: "I'm not sure what to do. I can't make much sense out of these articles and I really don't know how to give a talk on this. There's just too much stuff here for fifteen minutes," how do you respond?

3. Think about Ernie for a second. If you run into him next semester and he claims to be completely lost in his differential equations course, what do you say?

4. Think about the process you go through in preparing a talk based on a research mathematics paper you have recently written. How do you design the talk? Does giving the talk improve your paper? If so, how? If not, why not?

Coda F — A Sample Major Project

In the previous chapters, we've generally outlined how to design and use major writing projects in mathematics courses. Here we briefly discuss one particular example from an elementary geometry class in order to illustrate how these various ideas knit together. This assignment deepens the students' understanding of *geodesic* and is most appropriately used after the class has explored geodesics in various geometries (perhaps the sphere, torus, and/or hyperbolic plane). The topic is conical geometry, an important part of mathematics due to the use of "orbihedral structures" in geometry and topology.

This particular example involves a group of four students working on a "directed research project." Such an assignment balances good technical expository skills with low-level mathematical research. The students are guided to their results by a carefully designed worksheet and one or two conferences with their instructor. Generally the first questions on the worksheet are designed to help the students formulate good definitions and useful methods and the later questions indicate some of the peculiarities of their project. The conferences with the instructor help the students formulate and focus their research. In a sense, the conferences help with the prewriting. Student research projects should be structured so as to encourage the student researchers to:

- invent their own terminology;
- make conjectures about their topic;
- organize the body of knowledge it discovers.

If the students are at an advanced level, we would probably also require that they

- present their work in the theorem-example-lemma-proof format.

Our particular example (actually it's an amalgamation of two separate uses of similar assignments) began with the following worksheet.

Conical Project *

A *cone of angle* θ is the geometric object formed by taking a sector of angle θ in the Euclidean plane and identifying its two boundary rays. Recall that we have defined a *geodesic* in a geometry to be a path which is locally length minimizing. In this project you will be asked to explore and explain the behaviour of geodesics on cones. Your answers may well be different for cones with different angles.

1. Find a dozen examples of geodesics on various cones. While all of your geodesics have to be locally length minimizing, are they all minimal length paths between their endpoints? Did you try to find geodesics that go through the cone point?

---

* This project was developed from similar materials in [H].

2. Below is a cone formed from a sector with an angle of 90°. Is the indicated path a geodesic? Why or why not?

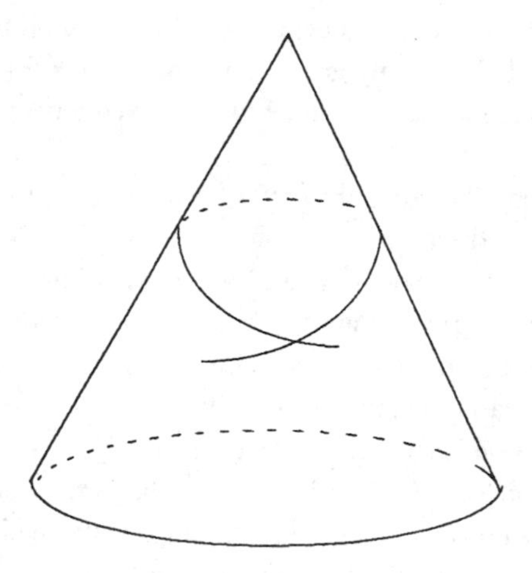

FIGURE 15
Is this a geodesic?

3. Can you construct other geodesics which self-intersect? Do geodesics on the sphere or torus self-intersect?

4. Prove: On a cone, every pair of points can be joined by a geodesic.

5. When studying the geometry of geodesics on the torus, we found it helpful to look at the Euclidean plane chopped up into squares, each square being identified with our torus. Similarly, if you are studying the geodesics on a 90° cone, you might want to look at four 90° sectors taped together to form the Euclidean plane. Use this idea to give a complete description of *all* geodesics between two arbitrary points x and y. Does this picture help you explain the geodesic in question 2?

6. Repeat question 5 using various different cone angles. For example, what changes if $\theta = 60°$ or 45°? What if your angle doesn't evenly divide 360°? What can you say about geodesics on a cone of angle 50°?

Spend the next week working on these questions, and then come by my office to discuss your progress. Be prepared to indicate one or two specific aspects of conical geometry you would like to explore in your paper.

After working on these questions for a week, the group was well prepared for their conference. They were quite excited about the project, especially because they had succeeded in constructing geodesics with large numbers of self-intersection points. They could show that a cone of angle 90° has geodesics with one self-intersection, a cone of angle 45° has geodesics with two self-intersections, and, if the angle is 30°, you can get three self-intersections. "We think there's a pattern here, but we're not quite sure how to prove it." A bit of pressing showed that the pattern they saw was: a cone with angle $\frac{360}{4n}$ contains geodesics with n intersections. I was quite impressed by their progress.

Since the students were intrigued, and the project would be challenging, I encouraged them to focus on using the "covering space theory" alluded to in question 5 to pursue their conjecture, and to begin considering other angles, such as $60°$ and $72°$. After an hour-long discussion of their findings and fledgling proofs, the students and I began to outline one possible paper they could write. In addition to clarifying my goals and expectations for their paper, this discussion made it quite clear that the work they had done was an excellent start, but that there was quite a bit of hard work ahead.

One week later the students returned to my office, having proved their $4n$-conjecture: the maximum number of self-intersections on a cone of angle $\frac{360°}{4n}$ is n. Further, as part of their proof they gave explicit instructions on how to construct such maximally intersecting geodesics. We spent a half–hour discussing how they had written their proof, improving their prose, and debating different ways to incorporate the proof into their paper. Eventually we agreed that their paper would have the motivating examples come before the formal statement of their result, then demonstrate a formal proof that their examples had the maximum number of intersections.

The group was still a bit vague about the $4n+2$ case, or any odd case, but time was short, so we decided that writing out a few particular examples in detail would suffice. In the last couple minutes of the conference we talked about how best to introduce their material. Over the weekend they produced a well constructed paper focused on the $4n$ case, with concluding comments and questions about the remaining cases. After one more draft they had produced an excellent manuscript with illuminating examples and complete proofs ... and these were sophomores!

Let's back up for one moment and look at the larger picture. How was this cone theory project integrated into the class as a whole? This project was just one of three in the class, the other two being on Thurston paper (a piecewise Euclidean model of the hyperbolic plane) and the three-sphere. In the class meeting after their final papers were turned in, each group stood up for 25 minutes and presented their findings to the rest of the class and distributed copies of the three papers. Each presentation and paper was the class's introduction to the main topics we would pursue in the remaining weeks of the semester: the geometry of surfaces with singularities, hyperbolic geometry, and three-dimensional geometry.

Each group was fully aware of the central role their work played in the syllabus, and every student was aware that they would have homework based on the material presented by their classmates. (For example: What's the maximum number of self-intersections that can occur on a geodesic in a cone of angle $\frac{360°}{7}$?) Quite often students would quote their classmates in their homework and examinations. "As Susan pointed out in the presentation on conical geometry ...," or "Similar to the main theorem in Thurston paper, in the hyperbolic plane triangles have interior angle sums less than $180°$."

Not only were the groups motivated by the desire for good grades and their interest in the topic, in a very real sense their classmates and the course depended on them to produce good work. Their work was so good that two of these groups went on to present their work at the Moravian College Undergraduate Mathematics Conference. One of the students who worked on the conical geometry project participated in an REU program studying geodesics on punctured surfaces, and the Thurston paper group eventually produced a classroom capsule which they submitted to the *College Mathematics Journal* [B+].

Narrating Mathematics

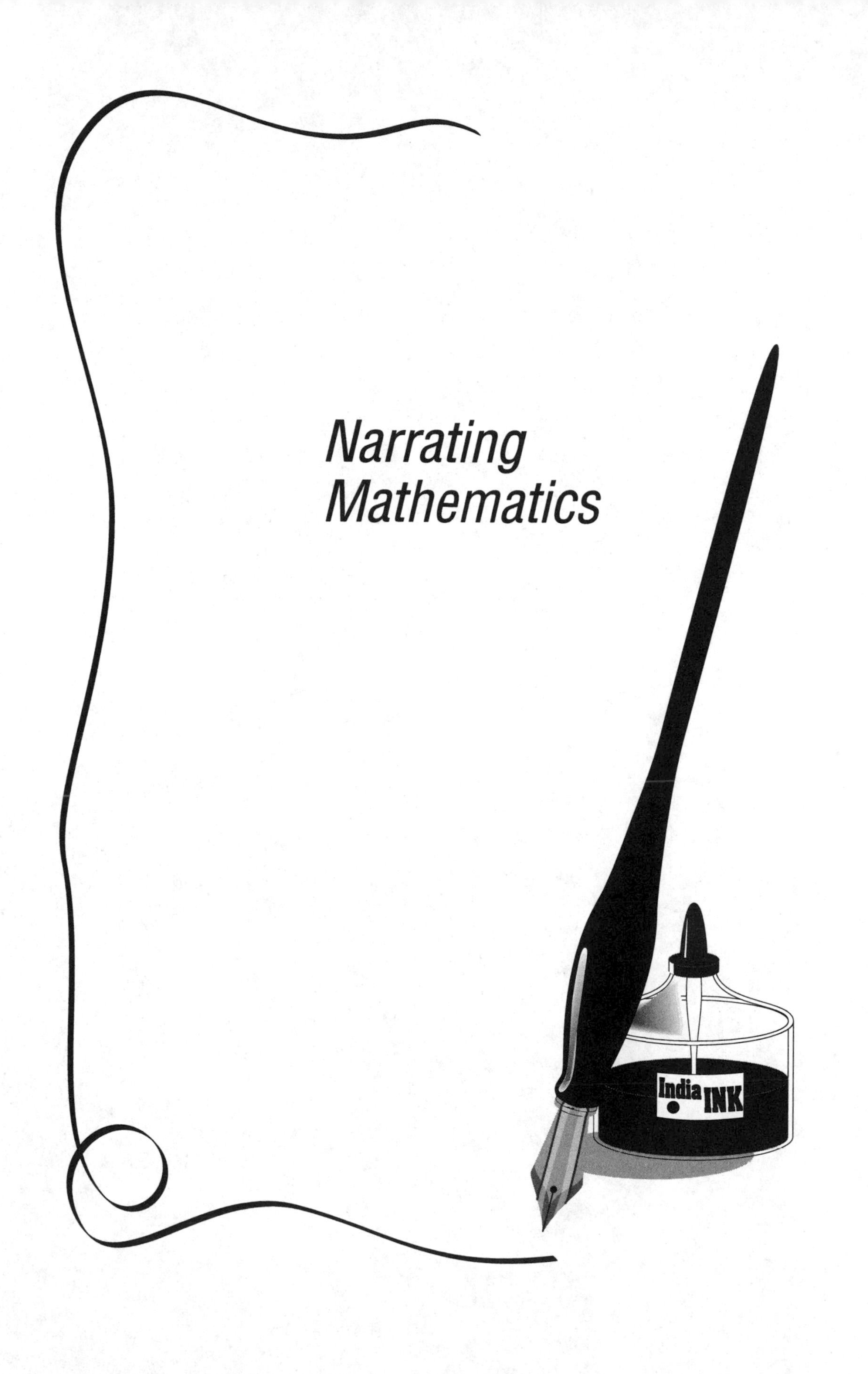

Prelude — "Learning of Limits"

A few years ago, Tom decided it was time to have a course called Mathematical Exposition, designed for advanced undergraduate mathematics and science majors who would write papers and give talks on mathematically related topics to varied audiences. Assignments would begin simply: "Give a five-minute talk on your favorite theorem. Why is it a 'favorite'?" "Interview a scientist or mathematician you think interesting. Find out a little about his or her background and the area in which he or she does research."

After having gotten the students used to easy writing and speaking assignments, Tom asked them to go on to "a topic which is central to a course you are currently taking or have just finished. Why is the topic important? Find some ideas that are still confusing or unclear about the topic and discuss them."

The exposition course heads toward a major project, which is a subject mutually agreed upon by Tom and the student. Recent papers have been on such topics as cryptanalysis, curvature of surfaces, linear programming, nonstandard analysis, and American mathematics in the 1930s and 1940s.

Tom's impulse for designing the course came from a number of factors:

Mathematics majors almost never get to speak in a classroom setting, since lecturing is the main form of mathematics instruction; hence, it is difficult to assess how well information is being processed. (For graphic proof of this point, read the next chapter.)

Many majors are not going on to graduate school—at least not in mathematics—but rather into business or computer situations. These students will need to be able to explain technical information to co-workers and clients, verbally and in memos. This skill needs to be acquired somewhere; where better than in a setting in which technical expertise can be carefully assessed.

For those students who are going on to graduate school in mathematics or science, the career goal is sometimes teaching, sometimes research. In either case, exposition will be a significant component of the student's future.

Now, for a real-world case study of what actually happened in class...

14

Learning of Limits; Limits of Learning

A Case Study on the Impulse toward Narrative in Mathematics*

For some years now, I have been incorporating expository methods into my mathematics courses.

To get the students quickly involved in the central theme of the course, I asked them each to begin by giving a five-minute talk on a "favorite equation," and to turn in their notes at the end of the discussion. The first two talks proceeded in an entirely expected manner, but suddenly everything changed.

The third speaker said that he wanted to discuss the "theorem"

$$dy/dx = \lim[(f(x_1) - f(x_0))/(x_1 - x_0)].$$

The third time he referred to the definition of the derivative as a theorem, I asked him and the class the difference between a definition, a theorem, and, "... for that matter, an axiom, a hypothesis, a proof, etc." There was obvious consternation in the room as more and more of the students realized that they didn't know, or couldn't articulate, the differences. Storing the information for future discussion, I told the speaker, "... but I interrupted your talk; continue on with your discussion of the derivative." So we began to take the derivative of $y = x^2$ by the limit process.

By now aware that I might be dealing with something that the students needed to study more, I waited until the speaker divided $(x_1 - x_0)$ by itself, and then I asked how he could do that. "Can you divide before you take the limit? Aren't you dividing zero by zero?" Now the entire class plowed directly into the discussion. For 20 or 30 minutes, they tried to figure out, with some prodding from each other and me, whether or not they were dividing zero by zero; why they could "do algebra before taking the limit"; what was the difference between "limit" and "continuity"; what did that have to do with the notion of "derivative"; and so forth. I did not help them; in fact, I added to the muddle by bringing out more and more objections to each argument.

At the same time as I was confusing the class, I was thinking through some serious pedagogical problems: "Do I jump up, grab chalk, and say, 'Look! You're supposed to know this!'?—Of course not; these students had had four years of that, and it didn't work. O.K. then,

* This chapter is a revision of [R+].

this *is* a writing class; how about assigning a paper? Nope, they'd either copy from a text, or write all kinds of misinformation.

With about three minutes to go in the class, I said, "Wait a minute. Let's summarize the questions we have. You tell me what they are, and I'll put them on the board." Once we had our list of questions, I told them, "Class is over; *but*, we'll come back to this topic."

The next class continued with "favorite equations" presentations; but at the end of class, I said, "I'll bet some of you have been looking through books for the word 'limit.' " On cue, one student pulled out a calculus text, responding, "Yeah, I read the definition — but I still don't get it." "All right! For your next assignment, I want each person to find out two pieces of information about the word 'limit'; if you look it up in a book, reference that book; bring in what you discover, and we'll start discussing what it means." Meanwhile, I handed out what I originally considered to be my next "real assignment" in the course.

Students brought in quotes, leading to a discussion of everything from the epsilon–delta definition of limit, through some historical comments about Archimedes to a statement of the mean value theorem, ("I thought I'd bring in the statement of an important theorem using the derivative," explained the student), to the definition of convergence of sequences in metrizable spaces ("That's what we're studying now.") We "reviewed" all these concepts, trying to find how they did, or did not, fit together, while simultaneously attempting to understand their relationship to other concepts which came up in our conversation.

A few classes later, I returned to the topic by guiding another discussion; this time, about the history of the word "limit."

One evening toward the middle of October, I noticed that about six of my students had addressed ideas in their "regular assignments" which had something to do with the limit concept. The next class, I said, "You know, we could give a talk at the Math Club. John could tell, if he is willing, how we entered into this topic; then Max and Julie could talk (five minutes each) about Eudoxus and Archimedes, which they've already written on; then Dan on Fourier and pathological examples; and Alex on nonstandard analysis." The students were obviously very interested, but still, I said, "You have to think about it for a day. Let me know next time." The next day, they, of course, agreed to the talk.

Two classes later, we gave each other two-minute summaries of our "five-minute talks," so that we would have a narrative line to our story about limits. The next day, I brought in three shopping bags full of books: history of mathematics texts, analysis books, mathematical dictionaries, and encyclopedias. We spent 75 minutes at the table, combing through the texts, looking for leads for ourselves and each other; rejecting some ideas, adding a few that came up in discussion. Then, on Tuesday, November 8, we got together one more time to make overhead transparencies and to practice our talks on each other.

On November 9, the students gave a very well received nine-person talk, "The Evolution of the Concept of Limit," to the Cornell Math Club.

Now let me make a few pedagogical comments about what the students taught me from this experience.

First: There is a real difference between having been taught something and internalizing that same thing. Every one of these students had been shown epsilon-delta proofs, but none had really absorbed them. (Of course, we faculty all know this—but it's amazing how often we need to be reminded.)

Second: Reflection is essential to the learning process. An assignment which starts: "Go outside and walk around the quad twice, and just think about . . . " is as meaningful as a problem set. In fact, it may be more meaningful, since it's so seldom assigned in mathematics.

Third: Prewriting, which can include discussion, note-taking, in-class exercises, "just thinking," is an important prologue to "real writing and thinking," and certainly to real mastery of a topic.

Fourth: All mathematicians, even great ones, have phases where they just "don't get it," and these phases are important to the learning process. We must respect the confusion our students have about those matters which may have become trivial to us; and we must find ways to bring these matters to the fore.

And finally: Mathematics is no different from the rest of experience; it is a topic which we store in our heads as a narrative—a story about what we know, what we don't know and what we wish to know. It's a story we must share with other people to see if we've got it right, to fill in the gaps, to make it grow. We must encourage our students, and ourselves, to share this narrative that we each have constructed and will construct. "Storying, encountering the world, and understanding it contextually, by shaping ideas, facts, experience itself into stories," says Gordon Wells, [Wg] "is one of the most fundamental means of making meaning: as such, it is an activity that pervades all learning."

15

A Rationale for Writing

A fellow faculty member asks why you use writing so much in your classroom. "You're never gonna convince me that it has anything to do with mathematics, you know... ."

When we attempt to "do mathematics," at any level, we are immediately confronted with the question of how we communicate our ideas. Well, whether we speak concepts or write them, we are using language. That's the short, simple answer to the "fellow faculty member" quoted above—we can't get away from language.

But there is more to be said about the challenge posed. Most writers of mathematics, be they students, faculty, or especially, researchers, would see themselves as writing *expository* prose, that kind of writing which is pure description or depiction, and which is informed by mere "following the recipe" ("Next take the derivative. Set it to zero. The answer is..."). Surely, some prose, in most any manuscript, is of this type ("He was a large man, with green eyes, and..."). But, there is another kind of writing going on in even the most abstract, research-oriented prose—a kind of writing that we often choose not to recognize, even though it is definitely there. For instance, when a research paper begins with, "In 1949, Jones wrote the first paper on...," the writer is probably starting to inform the reader of the place of that writer's paper in the historical development of an entire field of research. If, instead, the writer chooses to tell us that "In 1963, Smith was finally able to prove the Jones conjecture...," this time we are possibly being told about the end of a certain kind of era; or maybe we are going to hear about all the wonderful uses to which Smith's solution has been put in the last three decades. If the paper goes on to talk about "... the fact that Smith was able to lay to rest the last great problem in the field...," it would, implicitly or otherwise, make a political and cultural comment about the field of research, as well as the people who might continue to work on the topic.

But our goal here is not to analyze individual mathematical manuscripts in depth; suffice it to say that, in the interstices of papers, between the theorems and examples, are comments and statements of a narrative form—comments which inform us about the historical and social implications of the topic at hand. These comments and statements tell us a *story* of the paper, of the proof.

(Ironically, on the day after we wrote the above, we went to a topology seminar at which the speaker said such things as "If I had known of the analogy between the van Kampen theory and what I was working on, I would have saved myself two months," and "Essentially all the theory I'm going to tell you is embedded in the $PSL_2(Z)$-example; just think about that one example." The speaker was offering us a picture and telling us an anecdote to "hang our hat on" as we listened to his proofs. But then again, maybe we were receptive to these comments of the speaker at this time because we had been writing the thoughts in this chapter.)

It is the same, and more so, when *students* write about mathematics; they are placing the subject in a context which makes sense to them. If they are going into the field, the narrative they learn is the story they will carry into their subsequent courses to inform them as to why they need to know the definitions, the theorems, the proof techniques. If they are not going to continue in mathematics, they will probably forget the body of material they have studied; but, if they have written about the course, they will be much more likely to hold that narrative in their memory as their record of what that area of mathematics is about.

So, mathematics is embedded in language, and we must learn how to use that language well to express our ideas, to make them precise, to exhibit their worth to the outside audience—whether that audience is the instructor, or other students, educated mathematicians everywhere, or the world at large. It behooves us to learn to communicate our ideas well.

This is not to say that there aren't individuals out there who don't argue forcefully for the opposite of what we've just said. Here, for instance, is a quote from an article in the *American Mathematical Monthly* by David Gries and Fred B. Schneider [GSc]:

> Teaching mathematics through informalism is like driving in a fog. One sees dim figures in the distance, and every once in a while some of them suddenly appear clearly, but usually everything is veiled and mysterious. It's dangerous to drive in the fog, especially in a strange territory, and one must drive slowly. Even so, one may not always be sure where one is. Teaching rigor and precision, provided it is done without the veil of complexity interfering, burns away the fog, leaving everything crisp and clear and making it possible to drive faster and to enter uncharted lands.

It is interesting to see Gries and Schneider using analogy and metaphor, the very language they denigrate in mathematics, to defend their assertion that "mathematics and rigorous thinking can be taught more effectively by first teaching the design of rigorous proofs using a formal logic." We have no argument with most of the above statement; perhaps all we would do, if we were writing these passages, is remove the word "first" from the previous sentence.

There is a place for rigor in every kind of learning. All of us need to sharpen our thinking, and when we are challenged we should have to become more precise about what it is we are saying, how we are saying it, whether we are marshalling good arguments, and the like. Yet, at the same time, we also need to have a context for what we are doing. Why is the topic important? What are some of the examples that inform it? What is its history? These are all questions which can be most quickly and easily put into clear, intelligible prose for the neophyte that we all were once, and will be again the next time a seminar speaker comes to town.

Students always provide a good laboratory for our theories. Just when we think we can generalize about "what proof means," they teach us otherwise. Just one example may suffice to make this point.

Suppose we are teaching honors calculus. We decide to find out on exam two who's "getting it" and who isn't. Question five on the test is:

Prove that cubic polynomials can have no more than two maxima or minima.

Maybe the best answer we can expect in this context is the following:

Since

$$y = ax^3 + bx^2 + cx + d$$

$$= 3ax^2 + 2bx + c = 0$$

$$x = \frac{-2b \pm \sqrt{4b^2 - 12ac}}{6a}$$

So there are two roots.

FIGURE 16

We see immediately that there are problems with this solution: the student seems not to know the difference between "implies" and "equals"; nowhere does he say anything about taking a derivative (although he has obviously done so); multiple roots are not considered; the relationship between a function and its derivative is not explicit, punctuation is nonexistent. While we might give the student eight points out of 10, say, we are left with the queasy feeling that he isn't in command of what he is saying.

Suppose, instead, we ask: Write a *short* note explaining why cubic equations should have two or fewer maxima or minima.

One answer is shown in Figure 17.

Cubic equations have degree three.
They look like:

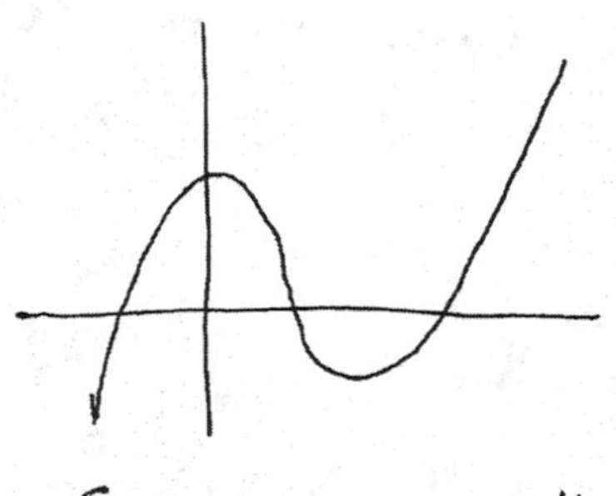

So we can see that there is only one max and min.

FIGURE 17

The same answer with our response is shown in Figure 18. Now we are engaging the student in a dialog.

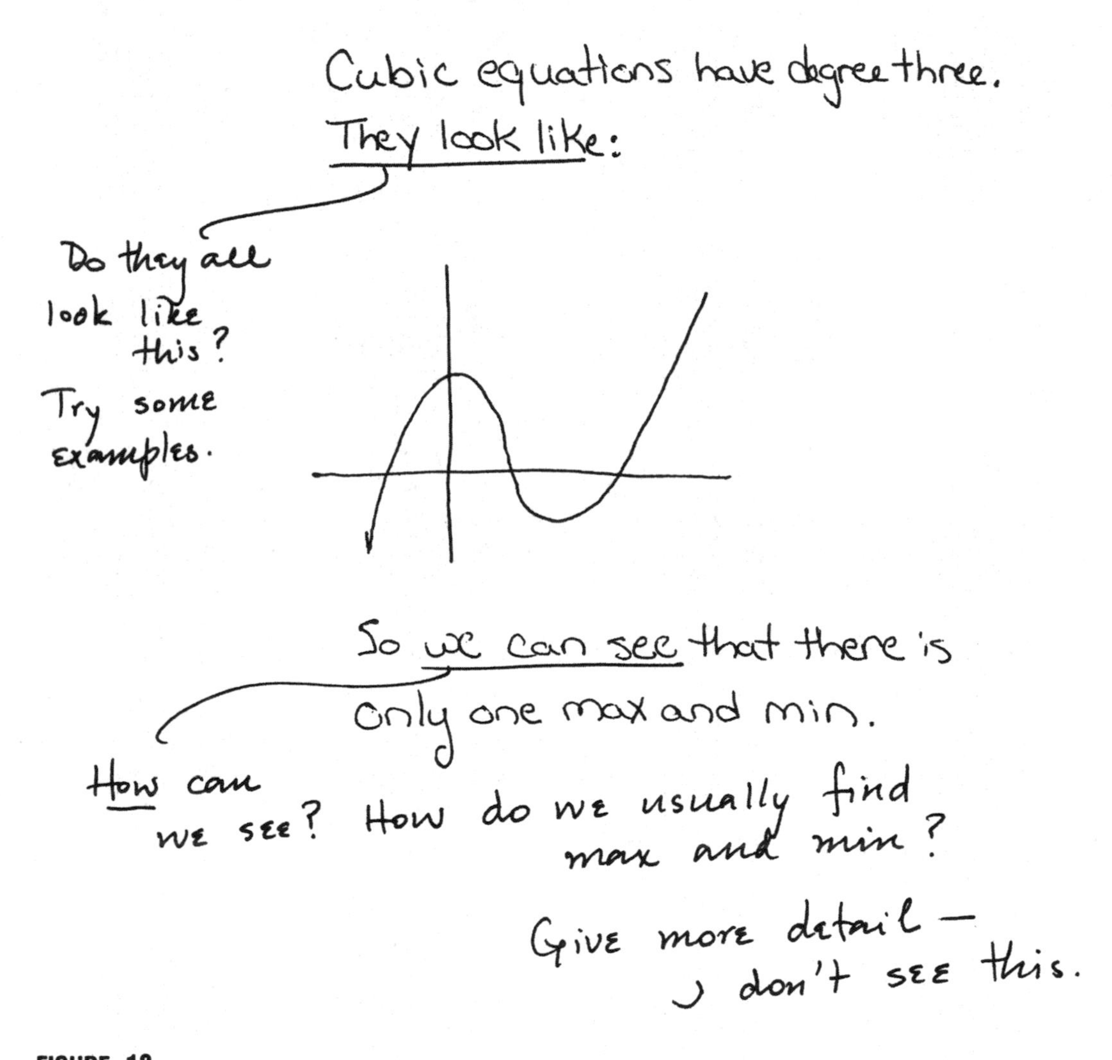

FIGURE 18

His response, with our comments is shown in Figure 19 on the next page. The second effort, while not yet a "proof," shows that the student is moving toward his own awareness of the use of the derivative in the solution. What he has not yet found is the difference between a specific example and a generalization. That starts to show up in the next draft (Figure 20).

All cubics? Or just some? What's the most general cubic?

Cubic equations are like

$$y = x^3 - 5x^2 + 3x - 1$$

when we take the derivative, we get

Good. But why do you want to do this?

$$y' = 3x^2 - 10x + 3$$

This has max or min when

Are max-min the only possibilities?

$$3x^2 - 10x + 3 = 0,$$

so $x = 3$ or $x = 1/3$.

What kinds of roots do cubics have?

The graph is

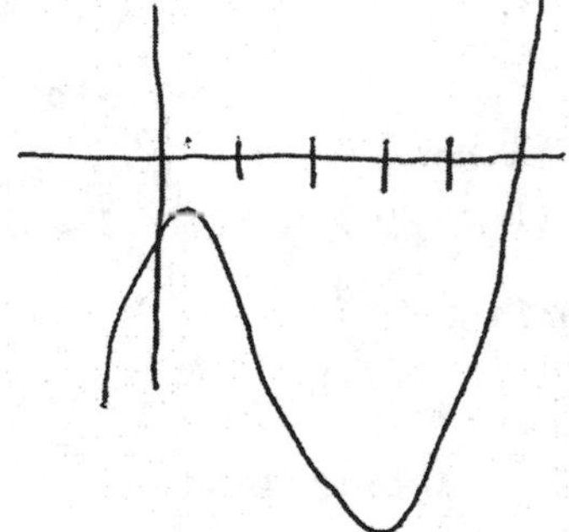

So the derivative says there is one max and one min, which is two, which is what we wanted to show.

For this one equation? Or for all cubics?

Try once more; getting close to a proof. You might compare your example with $y = x^3 - 3x^2 + 3x - 1$, and with $y = ax^3 + bx^2 + cx + d$.

FIGURE 19

A cubic equation has the form

$$y = ax^3 + bx^2 + cx + d$$

when we take the derivative, it tells us how many max, min, and p.o.i we have. For the cubic equation, we get

$$y' = 3ax^2 + 2bx + c,$$

a second-degree equation. This equation always has 2 roots, but sometimes they are equal, like in $y = (x-1)^2$ where roots are 1 and 1. Since the roots tell you how many critical points, this has two.

Good. P.o.i. are also important here. Do you know why?

But isn't this just one root which occurs twice? So, what does this say about critical points?

This is close to a proof, but slightly muddled. Look at the general case $y = ax^3 + bx^2 + cx + d$, then consider the possible roots of the derivative function. Finally, compare the $y = (x-1)^3$ example and the $y = x^3 - 5x^2 + 3x - 1$ case you did before.

FIGURE 20

Notice this last comment; through narrative and discovery, the student has not only come up with a proof of the original assertion, he also has a further conjecture, and a pretty good idea of how he would prove it.

At this point, a skeptical reader may claim that we are feeding the answers to the students, but we disagree—at every stage in the process, it is *the student* who is discovering the elements of the proof; we are simply aiding the student in moving in the proper direction. Compare this with the first student, who is hoping to have told us what *we* already know.

Let us skip to the final draft of the student's paper (Figures 21a and b).

Cubic equations are of the form

$$y = ax^3 + bx^2 + cx + d.$$

By analyzing the derivative, we can find out how many times it turns around. Since

$$y' = 3ax^2 + 2bx + c$$

is a quadratic equation, it has roots of the form

$$\frac{-2b \pm \sqrt{4b^2 - 12ac}}{6a}.$$

These roots indicate where possible max, min and points of inflection will be. There could be two real roots, in which case there will be at most two critical points. There could also be one root of multiplicity two, giving at most one critical point. A third possibility is complex roots only, giving no critical points. In any case, there are no more than two critical points (max, min, p.o.i.), so no more than two max and min, which is what we needed to show.

FIGURE 21a

Here are two examples.

For $y = x^3 - 5x^2 + 3x - 1$,

$y' = 3x^2 - 10x + 3$

So $x = 3$ or $x = 1/3$. The graph is

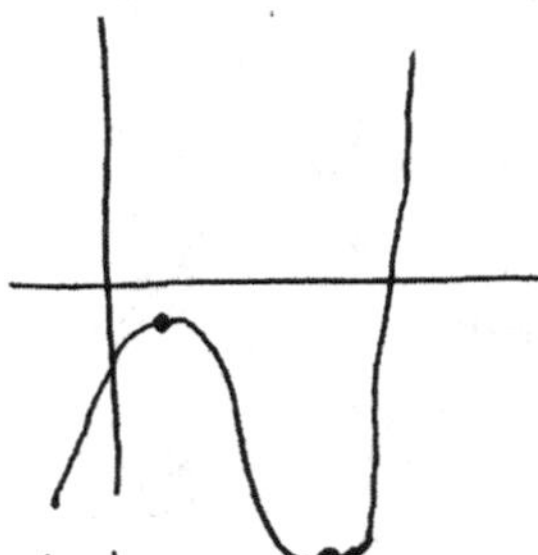

with max and min marked.

This is the standard case.

The other example is

$y = (x-1)^3 = x^3 - 3x^2 + 3x - 1$

$y' = 3(x-1)^2$,

So $x = 1$ is a root of multiplicity 2.

The graph is ↓ and there are no max or min.

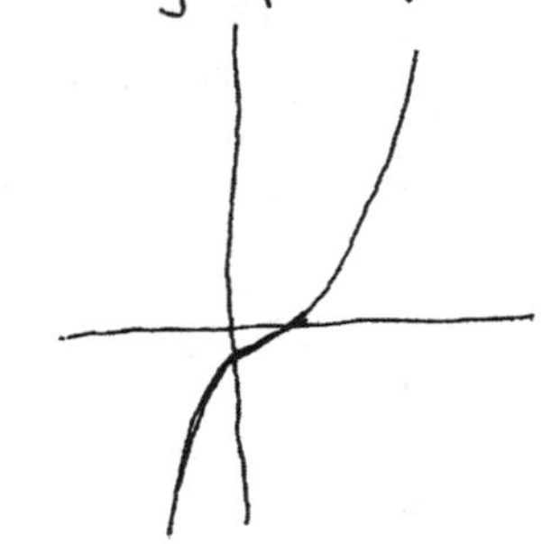

The reader can find a complex root example.

One last thing: I think this proof also works for higher degree polynomials, with the same kind of proof.

FIGURE 21b

In conclusion, let us say that each field has its narrative, which contains its history, its culture, its assumptions, its people, and its life. Mathematics is no different. Each mathematician constructs his or her own life in mathematics, and each student—and we are all students at some level—needs to find him—or herself within this narrative. Writing and speaking mathematics are central to learning and doing mathematics.

Bibliography

[Ab] Abbott, Edwin A. *Flatland.* New York: Dover, 1952.

[AA] Allen, Barbara, and Gwen Autin. "Maximizing the First Five Minutes of Class." *PRIMUS* **V**, 2 (1995) 139–148.

[Am] Ammons, A. R. *The Selected Poems, Expanded Edition.* New York: W. W. Norton, 1986.

[Ba] Bachelard, Gaston. *The Poetics of Space.* Tr. by M. Jolas. Boston: Beacon Press, 1969.

[Bn] Barnett, Janet Heine. "Assessing Student Understanding Through Writing." *PRIMUS* **VI**, 1 (1996) 77–86.

[Bs] Barr, Stephen. *Experiments in Topology.* New York: Dover, 1964.

[Bt] Barr, Thomas H. "Integrating Mathematical Ideas Through Reading, Writing and Speaking: A Senior Seminar in Mathematics and Computer Science." *PRIMUS* **V**, 1 (1995) 43–54.

[Be] Bell, E. T. *Men of Mathematics*. New York: Touchstone, 1986.

[BB] Bell, E. S., and R. N. Bell, "Writing and mathematical problem solving: Arguments in favor of synthesis." *School Science and Mathematics* **85** (1985) 210–221.

[B+] Benashaski, Meier, O'Brien, Reinheimer, and Skarbek. *"Introducing hyperbolicity via piecewise Euclidean geometry"*. Preprint.

[Bl] Bloom, B., ed. *Taxonomy of Educational Objectives.* New York: Longmans, Green, 1956.

[Bo] Boyer, Carl. *History of Mathematics.* Princeton, 1985.

[Bk] Buerk, Dorothy. "An Experience with Some Able Women Who Avoid Mathematics." *For the Learning of Mathematics* **3**, 2 (1982) 19–24.

[Bu] Burger, Dionys. *Sphereland.* New York: Barncs & Noble, 1983.

[By] Buyske, Steven G. "Student Communication in the Advanced Mathematics Classroom." *PRIMUS* **V**, 1 (1995) 23–32.

[Ch] Chaffee, J. *Thinking Critically.* Boston: Houghton Mifflin, 1994.

[CWS] Clark, David J., Andrew Waywood, and Max Stephens. "Probing the structure of mathematical writing." *Educational Studies in Mathematics* **25** (1993) 235–250.

[CV] Connolly, Paul, and Teresa Vilardi (eds). *Writing to Learn Mathematics and Science.* New York: Teachers College Press, 1989.

[Co] Countryman, Joan. *Writing to learn mathematics.* Portsmouth, NH: Heinemann, 1992.

[Cr] Crannell, Annalisa. "How to Grade 300 Mathematical Essays and Survive to Tell the Tale." *PRIMUS* **IV**, 3 (1994) 193–218.

[Cz] Czerwinski, Ralph. "A Writing Assignment in Abstract Algebra." *PRIMUS* **IV**, 2 (1994) 117–124.

[DH] Davis, Phillip, and Reuben Hersh. *The Mathematical Experience.* Boston: Birkhauser, 1981.

[Eg] Edgerton, S. Y., Jr. *The Heritage of Giotto's Geometry.* Ithaca: Cornell University Press, 1991.

[Ed] Edwards, C. H., Jr. *The Historical Development of the Calculus.* New York: Springer-Verlag, 1979.

[El 1] Elbow, Peter. *Writing without teachers.* New York: Oxford, 1973.

[Em 1] Emig, Janet. "Writing as a mode of learning." *College Composition and Communication* **28** (1977) 122–127.

[Em 2] Emig, Janet. *The Web of Meaning.* Portsmouth, NH: Boynton/Cook, 1983.

[Ev] Evans, C. S. "Writing to learn in math." *Language Arts* **61** (1984) 828–835.

[Fe] Fennel, F. "Diagnostic teaching, writing and mathematics." *Focus on Learning Problems in Mathematics* **13**, 3 (1991) 39–50.

[FY] Fulwiler, Toby, and Art Young, eds. *Language Connections: Writing and Reading across the Curriculum.* NCTE, 1982.

[Ge] Geeslin, W. E. "Using writing about mathematics as a teaching technique." *Mathematics Teacher* **70** (1977) 112–115.

[GSm] Gopen, George D., and David A. Smith. "What's an Assignment Like You Doing in a Course Like This?" *The College Mathematics Journal* **21**, 1 (1989) 2–19.

[Gy] Gray, Jeremy. *Ideas of Space.* New York: Clarendon Press, 1979.

[Gf] Griffiths, H. B. *Surfaces.* London: Cambridge, 1976.

[Gr] Greenburg, M. J. *Euclidean and Non–Euclidean Geometry.* San Francisco: Freeman, 1972.

[GSc] Gries, D., and F. B. Schneider. "Teaching Math More Effectively, Through Calculational Proofs." *The American Mathematical Monthly* **102** (1995) 691–697.

[H] Henderson, David. *Experiencing Geometry.* Englewood Cliffs: Prentice Hall, 1996.

[I] Isaacson, Zelda. *"'They look at you in absolute horror': Women Writing and Talking about Mathematics".* In *Gender and Mathematics: An International Perspective,* Leone Burton, ed, Cassell (1990) 20–28.

[J] Johnson, M. L. "Writing in mathematics classes: A valuable tool for learning." *Mathematics Teacher* **76**, 2 (1983) 117–119.

[La] Lakatos, I. *Proofs and Refutations, the Logic of Mathematical Discovery.* Cambridge, 1976.

[Lg] LeGere, Adele. "Collaboration and Writing in the Mathematics Classroom." *Mathematics Teacher* **84** (1991) 166–171.

[Me 1] Meier, John. "I'd like a function with a view of the lake." *PRIMUS* **IV**, 2 (1994) 185–191.

[Me 2] Meier, John. "Don't just tell me what your computer told you." *PRIMUS* **V**, 1 (1995) 1–22.

[MT] Meier, John, and Trisha Thorme. "Plotting Pots: Archaeological Exercises in Elementary Calculus." *PRIMUS* **VII**, 2 (1997) 129–140.

[Mt] Mett, C. L. "Writing as a learning device in calculus." *Mathematics Teacher* **80** (1989) 534–537.

[Mi 1] Miller, L. D. "Writing to Learn Mathematics." *Mathematics Teacher* **84** (1991) 516–521.

[Mi 2] Miller, L. D. "Teacher benefits from using impromptu writing prompts in algebra classes." *Journal for Research in Mathematics Education* **23** (1992) 329–340.

[MTW] Misner, Charles W., Kip S. Thorne, and John Archibald Wheeler. *Gravitation.* San Francisco: Freeman, 1973.

[Mo] Montgomery, Richard G. "A Two-Genie Strategy Helps Students Write Weekly Papers." *PRIMUS* **IV**, 4 (1994) 347–358.

[Ms] Morison, Samuel Eliot. *Admiral of the Ocean Sea.* Boston: Little, Brown, 1942.

[My] Myers, Nadine C. "Writing and Speaking to Learn Geometry." *PRIMUS* **I, (3)** (1991) 287–294.

[N] Nemerov, Howard. *Figures of Thought.* Boston: Godine, 1978.

[Pe] Petit, J-P. *Here's Looking at Euclid.* Providence: Janson, 1985.

[PC] Phillips, Eileen, and Sandra Crespo. "Developing Written Communication in Mathematics Through Penpal Letters." *For the Learning of Mathematics* **16**, 1 (1996) 15–22.

[Py] Polya, George. *Mathematics and Plausible Reasoning.* Princeton, 1954.

[Po] Porter, Gerald J. "Writing about Linear Algebra: Report on an Experiment." UME *Trends* **3**, 3 (1991) 1–3.

[Pr] Price, J. J. "Learning Mathematics through Writing: Some Guidelines." *College Mathematics Journal* **20** (1989) 393–401.

[Ri 1] Rishel, Thomas W. "The geometric metaphor: Writing and mathematics in the classroom." *PRIMUS* **I**, 2 (1991) 113–128.

[Ri 2] Rishel, Thomas W. "Assessment of writing in mathematics." *PRIMUS* **IV**, 1 (1994) 39–43.

[R+] Rishel, Thomas W. "Learning of Limits, Limits to Learning." *CMS Notes* 27, 6 (1995) 20–22.

[Ru] Rucker, Rudolf. *Geometry, Relativity and the Fourth Dimension.* New York: Dover, 1977.

[SST] Sabrio, D., S. Sabrio, and G. Tintera. "Writing to learn and learning to write mathematics: An experiment." *PRIMUS* **III**, 4 (1993) 419–429.

[SW] Schattschneider, Doris, and Wallace Walker. *M. C. Escher Kaleidocycles.* Corte Madiera: Pomegranate Artbooks, 1987.

[St] Sterrett, Andrew. *Using Writing to Teach Mathematics. MAA Notes* # 16, Washington: Mathematical Association of America, 1990.

[Sv] Stevens, Wallace. *The Collected Poems.* New York: Vintage Books, 1982.

[Sw 1] Stillwell, J. *Mathematics and Its History.* New York: Springer-Verlag, 1989.

[Sw 2] Stillwell, J. *Sources of Hyperbolic Geometry.* Providence: American Mathematical Society, 1996.

[Sk] Struik, D. *A Sourcebook in Mathematics, 1200–1800.* Princeton, 1969.

[V] Vallin, Robert W. "Letters to a Friend: Writing in Intermediate Algebra." *PRIMUS* **IV**, 2 (1994) 97-105.

[Wa 1] Waywood, Andrew. "Journal writing and learning mathematics." *For the Learning of Mathematics* **12**, 2 (1992) 34–43.

[Wa 2] Waywood, Andrew. "Informal writing-to-learn as a dimension of a student profile." *Educational Studies in Mathematics* **27** (1994) 321–340.

[We] Weeks, Jeffrey R. *The Shape of Space.* New York: Dekker, 1985.

[Wd] Wells, David. *The Penguin Dictionary of Curious and Interesting Geometry.* Penguin, 1991.

[Wg] Wells, Gordon. *The Meaning Makers: Children Learning Language to Learn.* Portsmouth, NH: Heinemann, 1986.

[YF] Young, Art, and Toby Fulwiler, eds. *Writing Across the Disciplines.* Upper Montclair, NJ: Boynton/Cook, 1986.

[Z] Zinsser, William. *Writing to Learn.* New York: Harper & Row, 1988.